Strange Brains and Genius

The Secret Lives of
ECCENTRIC SCIENTISTS
and MADMEN

D1004987

Previous Works by Clifford A. Pickover

The Loom of God: Mathematical Tapestries at the Edge of Time

The Alien IQ Test

Black Holes, A Traveler's Guide

Chaos in Wonderland: Visual Adventures in a Fractal World

Computers, Fractals, Chaos (in Japanese)

Computers, Pattern, Chaos, and Beauty

Computers and the Imagination

Future Health: Computers and Medicine in the 21st Century

Fractal Horizons: The Future Use of Fractals

Frontiers of Scientific Visualization (with Stu Tewksbury)

Keys to Infinity

Mazes for the Mind: Computers and the Unexpected

Mit den Augen des Computers (in German)

The Pattern Book: Fractals, Art, and Nature

Spider Legs (with Piers Anthony)

Spiral Symmetry (with Istvan Hargittai)

Time, A Traveler's Guide

Visions of the Future: Art, Technology, and Computing in the 21st Century

Visualizing Biological Information

Strange Brains and Genius

The Secret Lives of ECCENTRIC SCIENTISTS and MADMEN

CLIFFORD A. PICKOVER

QUILL
WILLIAM MORROW
NEW YORK

Copyright © 1998 by Clifford A. Pickover
First published in 1998 by Plenum Publishing Corporation.

All rights reserved. No part of this book may be reproduced or utilized in any form or by
any means, electronic or mechanical, including photocopying, recording, or by any
information storage or retrieval system, without permission in writing from the Publisher.
Inquiries should be addressed to Permissions Department, William Morrow and
Company, Inc., 1350 Avenue of the Americas, New York, N.Y. 10019.

It is the policy of William Morrow and Company, Inc., and its imprints and affiliates,
recognizing the importance of preserving what has been written, to print the books we
publish on acid-free paper, and we exert our best efforts to that end.

Library of Congress Cataloging-in-Publication Data

Pickover, Clifford A.
Strange brains and genius : the secret lives of eccentric
scientists and madmen / Clifford A. Pickover.
p.   cm.
Originally published: New York : Plenum Trade, c1998.
Includes bibliographical references and index.
ISBN 0-688-16894-9 (alk. paper)
1. Scientists—Psychology.   2. Genius.   3. Eccentrics and
eccentricities.   4. Scientists Biography.   I. Title.
Q147.P53   1999
509.2'2—dc21   99-20675   CIP

Printed in the United States of America

First Quill Edition 1999

1   2   3   4   5   6   7   8   9   10

www.williammorrow.com

This book is dedicated to the cracked,
for they shall let in the light.

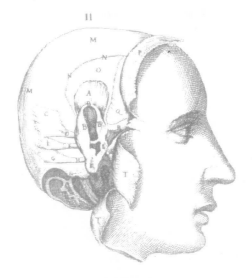

CONTENTS

CURIOSITY SMORGASBORD

FINALE

APPENDIXES

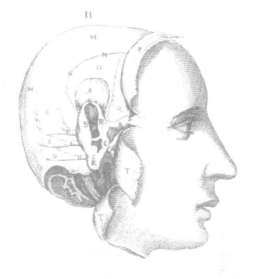

PREFACE

MEN HAVE CALLED ME MAD BUT THE QUESTION IS
NOT YET SETTLED, WHETHER MADNESS IS OR IS NOT
THE LOFTIEST INTELLIGENCE—WHETHER MUCH
THAT IS GLORIOUS—WHETHER ALL THAT IS
PROFOUND—DOES NOT SPRING FROM DISEASE OF
THOUGHT—FROM MODES OF MIND EXALTED AT THE
EXPENSE OF THE GENERAL INTELLECT.

EDGAR ALLAN POE

IF A MAN CANNOT KEEP PACE WITH HIS
COMPANIONS, PERHAPS IT IS BECAUSE HE HEARS A
DIFFERENT DRUMMER. LET HIM STEP TO THE MUSIC
HE HEARS HOWEVER MEASURED OR FARAWAY.

HENRY DAVID THOREAU

Weird Scientists—two words that conjure visions of eccentric researchers marching to drumbeats that no one else can hear. In repressive times, they've been persecuted, but in more enlightened eras these nonconformists have had the freedom to make great contributions to science and society. Are their minds like our own, or are they so different that these geniuses should be viewed as entirely different beings? What do geniuses have in common, and how can we foster their continued emergence? Is there a link between their obsessions and their creativity?

In this book, I'll avoid the well-known, influential eccentrics: the mad monks, kleptomaniac kings, and Wall Street sages. I am interested in *scientists* and *philosophers* with strange obsessions and compulsions. For example, early in *Strange Brains and Genius*, I discuss two geniuses who advanced our knowledge on electricity: Nikola Tesla, who had a fear of pearl earrings, and Oliver Heaviside, who replaced his furniture with granite blocks that sat in bare rooms like the furnishings of some stone-age giant. Most of the geniuses discussed in this book were celibate and never married, but through their energetic nonconformity they achieved greatness and changed our lives for the better. Interestingly, all the geniuses discussed in this book showed signs of brilliance in childhood. Most were ambitious and concerned with their reputations. Most had deep convictions about the correctness and importance of their own ideas. Very few had extraordinary parents. All had obsessive–compulsive tendencies. Most were born in Europe. Most had a parent who died early. However, not *all* scientific geniuses follow these rules. For example, Einstein didn't talk until he was three, and his humility is well known.

This book is organized into three parts. In the first part, I profile several geniuses who have obsessive–compulsive tendencies. Any discussion of temperament and genius is best served by examining one life in some depth, and none better illustrates the complex association than inventor Nikola Tesla with whom we start. At the end of Part I we take a break from the biographies of obsessive–compulsive geniuses to explore the obsessive–compulsive disorder itself. Individuals afflicted with obsessive–compulsive disorder are often compelled to commit repetitive acts that are apparently meaningless, such as persistent hand washing, counting, checking, and avoiding.

In the second part, I include a smorgasbord of short subjects ranging from IQ to the influence of the brain's structure on behavior; however, this book will not explain great scientists' behaviors in terms of brain anatomy. This would be impossible. For one thing, most of the great brains were not preserved and studied. However, Einstein's brain *has* been preserved for posterity, and, although he was not obsessive like the others in this book, I'll spend Chapter 12 describing his extraordinary brain convolutions.

In the third part, I discuss how individuals were selected for this book, summarize my thoughts on the association of genius and strangeness, and briefly describe the effect of other disorders such as bipolar disorder and temporal lobe epilepsy on creativity, religion, and even the alien abduction experience.

We may safely assume that there is a biological root to many of the unusual behaviors of great scientists. Recent theories suggest that obsessive–compulsive disorder, for instance, results from imbalances in the brain's chemistry. For example, afflicted individuals have brains that are depleted of an important chemical called serotonin. Today, drugs such as Anafranil are prescribed because they increase the amount of serotonin. Anafranil, known to biochemists by its chemical name clomipramine hydrochloride, is also helpful in treating obsessive–compulsive behavior in animals, especially in stopping dogs from licking their wounds so that the wounds can heal. Prozac (fluoxitine hydrochloride) also affects serotonin levels and can reduce obsessive–compulsive behavior as well as depression.

As I considered the lives of a number of especially creative scientists, inventors, and philosophers, I was impressed by the number of individuals who had curious deficiencies mixed with their more obvious talents. Although this book is not intended to be an academic analysis, we can still ponder the question: Can mental illness convey creative advantages to great scientists? Most scientists do not exhibit bizarre behaviors, and most people with mental disorders do not possess extraordinary creativity. However, a significantly large number of established *artists* have mood disorders such as bipolar disorder. (Bipolar disorder, also called manic depression, is a genetic illness characterized by states of depression and mania that may alternate cyclically. Bipolar disorder is closely related to major depressive, or unipolar, illness; in fact, the same criteria are used for the diagnosis of major depression as for the depressive phase of bipolar disorder.) In fact, it appears that both major depression and bipolar disorder can sometimes enhance the creativity of some people. So while we cannot say that the neurotic behavior of some great scientists causes their greatness, it likely plays a role.

Great eccentrics have intrigued historians for centuries. For example, *Jeremy Bentham*, the British philosopher who promoted the idea "the greatest good for the greatest number of people," fell in love with rats. He also advised rich people to plant embalmed corpses of their ancestors upright and above ground along stately drives. *Joseph Nollekens,* the 18th-century British sculptor, loved to eat the gristle and fat found on the butcher's floor as well as rancid butter. *Charles Kay Ogden,* Britain's most brilliant linguist, was a claustrophiliac—he loved to be shut in small places. Catholic naturalist taxidermist *Charles Waterton* turned animal corpses into effigies of famed Protestants, and after his wife died, he never slept in bed again, preferring the floor. Occasionally he would hide behind couches and attack guests like a dog, chewing at their ankles.

These are just a few samples of what I call "strange brains." Many of these great minds have had the compulsion to oddness, and I could go on and list hundreds of artists, musicians, and industrialists. But now it is time prepare yourself for the main subject of this book: the influential scientists, inventors, and philosophers. In exploring these geniuses, we explore ourselves.

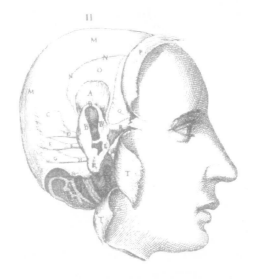

ACKNOWLEDGMENTS

THERE IS A THEORY THAT CREATIVITY ARISES WHEN INDIVIDUALS ARE OUT OF SYNC WITH THEIR ENVIRONMENT. TO PUT IT SIMPLY, PEOPLE WHO FIT IN WITH THEIR COMMUNITIES HAVE INSUFFICIENT MOTIVATION TO RISK THEIR PSYCHES IN CREATING SOMETHING TRULY NEW, WHILE THOSE WHO ARE OUT OF SYNC ARE DRIVEN BY THE CONSTANT NEED TO PROVE THEIR WORTH. THEY HAVE LESS TO LOSE AND MORE TO GAIN.

GARY TAUBES

THAT SO FEW NOW DARE TO BE ECCENTRIC, MARKS THE CHIEF DANGER OF OUR TIME.

JOHN STUART MILL
19TH-C ENGLISH PHILOSOPHER

thank Clay Fried and Yvonne Twomey for useful advice, Dr. Paul Hartal for the frontispiece figure of Einstein, and Trudy Myrrh Reagan for her brain/people illustration. I thank readers of Wisdom Forum and Reading Forum, two computer bulletin boards, for useful comments regarding portions of this book.

Some of the facts about genius in Chapter 16 come from A. Smith's *The Mind* (Viking). Some of the late 1800s drawings in Chapter 1 come from various copyright-free archives and L. de Vries' *Victorian Inventions* (American Heritage Press). This book's dedication and ending quotation are derived from a quotation by an anonymous eccentric in *Eccentrics, the Scientific Investigation* by D. Weeks and K. Ward (Stirling University Press).

The following books provide useful, general information on eccentric people:

- Editors of Time-Life Books (1992) *Odd and Eccentric People*. Time-Life Books: Alexandria, Virginia.
- LaPlante, E. (1993) *Seized*. HarperCollins: New York.
- Michell, J. (1984) *Eccentric Lives and Peculiar Notions*. Citadel Press: Secaucus, New Jersey.
- Weeks, D. and Ward, K. (1988) *Eccentrics, the Scientific Investigation*. Stirling University Press: East Kilbridge, Scotland.
- Weeks, D. and James, J. (1994) *Eccentrics*. Weidenfeld and Nicolson: East Kilbridge, Scotland.

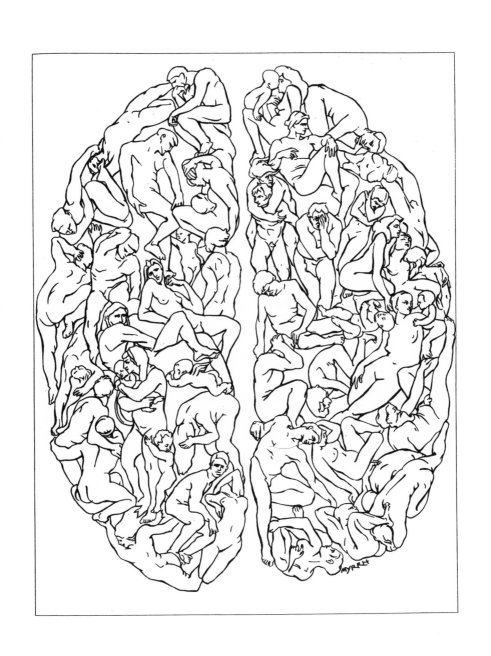

THROUGH A SCIENCE OR AN ARTFORM—THROUGH CREATIVITY—
THE INDIVIDUAL GENIUS SEEMS TO LIVE AT THE EXHILARATING EDGE
OF WHAT IT MEANS TO HAVE OUR HUMAN MIND.

JOHN BRIGGS

THE BRAIN IS A THREE POUND MASS YOU CAN HOLD IN YOUR HAND
THAT CAN CONCEIVE OF A UNIVERSE
A HUNDRED-BILLION LIGHT-YEARS ACROSS.

MARIAN DIAMOND

NO GREAT GENIUS IS WITHOUT SOME MIXTURE OF INSANITY.

ARISTOTLE (AS REPORTED BY SENECA)

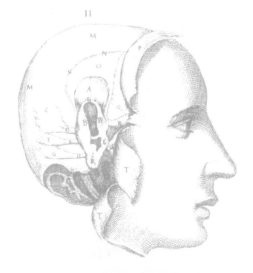

INTRODUCTION

TRUE ARTISTS AND TRUE SCIENTISTS HAVE FIRM
CONFIDENCE IN THEMSELVES. THIS CONFIDENCE IS
AN EXPRESSION OF INNER STRENGTH WHICH
ALLOWS THEM TO SPEAK OUT, SECURE IN THE
KNOWLEDGE THAT, APPEARANCES TO THE CONTRARY,
IT IS THE WORLD THAT IS CONFUSED AND NOT THEY.
THE FIRST MAN TO SEE AN ILLUSION BY WHICH MEN
HAVE FLOURISHED FOR CENTURIES SURELY STANDS
IN A LONELY PLACE. IN THAT MOMENT OF INSIGHT
HE, AND HE ALONE, SEES THE OBVIOUS WHICH TO
THE UNINITIATED (THE REST OF THE WORLD) YET
APPEARS AS NONSENSE OR, WORSE, AS
MADNESS OR HERESY.

GARY ZUKAV
THE DANCING WU LI MASTERS

IT IS, SO SAY HUMANS, THE MOST IMPORTANT THING
IN THE WORLD, BUT IT LOOKS AS INTERESTING AS
INTESTINES, AND INDEED WAS FREQUENTLY DRAWN
FORMERLY AS IF INTESTINAL, A TUBE FROM START
TO FINISH. OUR FOREFATHERS WERE MORE
INTRIGUED BY THE PULSING HEART, THE MOODY
SPLEEN, THE COLOR-CHANGING LIVER, THE
WANDERING AND PERISTALTIC GUT. EVEN URINE, IN
THEIR OPINION, HELD MORE EXCITEMENT
THAN THE BRAIN.

ANTHONY SMITH
THE MIND

Sometimes you are a brain-snatcher.

You imagine yourself the Chief Curator of a futuristic museum of brains. You walk down fluorescent corridors filled with gray, wrinkled brains stored in formalin-filled jars to prevent decay.

On your left are the brains of the brilliant writers, artists, and composers who had bipolar disorder (manic depression), a genetic illness characterized by states of depression and mania that may alternate cyclically: Sylvia Plath, Walt Whitman, Cole Porter, Anne Sexton, Vincent van Gogh, Gustav Mahler, John Berryman, Edgar Allan Poe, Virginia Woolf, Herman Hesse, Mark Rothko, Mark Twain, Charles Mingus, Tennessee Williams, Georgia O'Keeffe, and Ezra Pound. In one smaller bottle are some fragments of Ernest Hemingway's manic-depressive brain—all that is left after he shot a bullet through his skull.

You give a little tap on the jar marked "Poe." His cerebrum jiggles. *Never more, never more.* Genius and insanity are often entwined. You put Poe back in his place.

Today you are not interested in the artists and writers but in the strange brains of great scientists. Instead of having bipolar disorder, many great scientists in your collection were obsessive–compulsive—they felt compelled to commit meaningless repetitive acts such as excessive hand washing, collecting, or counting.

You walk a little further, wrinkling your nose at the strange chemical odors.

On your right are a few clear jars. You reach for the one marked "Isaac Newton," open it, and drag your fingers over his gray-white frontal lobes. Might there be remnants of his genius preserved in his neuronal networks:

the time he formulated the law of gravitation or studied the nature of light? Could some fossil of his hatred[1] toward his father and mother be buried within his brain's strata like an ancient ant trapped in amber? How could this great scientist have been such a suspicious, neurotic, tortured person? There were so few students going to hear Newton's lectures at Cambridge that he often read to the walls.

The brain: three pounds of soft matter that can take a split second of experience and freeze it forever in its cellular connections. A 100 billion nerve cells are the architecture of our experience. Recent studies have even shown that human talents are reflected in our brain structure. As just one example, consider the dendrites—tiny branches that convey signals to nerve cells. It turns out that machinists have more dendrites in certain areas of their brains than salesmen, who are less clever with their hands.

Is Newton still here in the wet organ draped by your palm? Could we reconstruct his memories? Would Newton approve such a breach of privacy?

You return Newton's jar and glance longingly at some of the other scientist brains in your possession: *Oliver Heaviside*, an eminent, brilliant Victorian mathematical physicist whose nails were always cherry pink; *Henry Cavendish*, one of the greatest scientists in British history who made discoveries in diverse fields of chemistry, electricity, and physics but who was so shy that he ordered his female servants to remain out of sight or be fired; *Sir Francis Galton*, distinguished British explorer, anthropologist, and eugenicist known for his pioneering studies of human intelligence, who once resolved to taste everything in the hospital pharmacy in alphabetical order. He got as far as "C" and swallowed some castor oil. Its laxative effects put an end to his gastronomical experiments.

Heaviside, Cavendish, and Galton are perhaps better preserved than Newton. Their brains are perfused with glycerol and frozen to −320 degrees Fahrenheit with liquid nitrogen. Your cryonicist friends refuse to give up hope that memories still reside in the brain cell interconnections and chemistry, much of which is preserved. Maybe they are right. After all, far back in the 50s, hamster brains were partially frozen and revived by British researcher Audrey Smith. If hamster brains can function after being frozen, why can't ours? In the 1960s, Japanese researcher Isamu Suda froze cat brains for a month and then thawed them. Some brain activity persisted.[1] Even as far back as 1891, Dr. Varlot, a surgeon at a major hospital in Paris, developed a method for covering people with a layer of metal in order to preserve them for eternity. This approach, however, probably did not appeal much to those hoping for eventual resurrection.[2]

1. Electroplating the dead.

But what if there is an afterlife? You bang on the giant thermos bottle containing Oliver Heaviside's brain, causing the brain to splash, sounding like a drunken fish. When he died Heaviside's brain was immediately frozen. Therefore, if there is an afterlife, he must have already experienced it by now. What would happen if his brain were revived?

You shake your head to change your direction of thoughts.

There is one gem missing from your collection: Nikola Tesla, a visionary genius, a great electronics inventor, a man disturbed by round objects, particularly the pearls in women's jewelry.

You press a time-travel button on your belt and are transported to the day of his funeral service: four o'clock, January 12, 1943. You huddle in the back of the Cathedral of St. John the Divine in New York City. A shiver runs along your spine.

There are over 2000 people. Honorary pallbearers include Dr. E. F. W. Alexanderson of General Electric, Dr. Harvey Rentschler of Westinghouse, W. H. Barton, curator of the Hayden Planetarium of the American Museum of Natural History, and Professor Edwin H. Armstrong. New York mayor Fiorello LaGuardia had just read a moving eulogy to Tesla over radio station WNYC. The President and Mrs. Roosevelt express the country's gratitude for Tesla's scientific contributions. Three U.S. Nobel prize winners in physics—Robert Millikan, Arthur Compton, and James Franck—participate in the eulogy calling Tesla "one of the outstanding intellects of the world who paved the way for many of the important technological developments of modern times." Author Louis Adamic eulogizes, "Tesla lives in his achievement, which is great, almost beyond calculation, and an integral part of our civilization, our daily lives, our current war effort. His life is a triumph."

You press your time-travel button again, this time to stop time. You are still in the Cathedral of St. John the Divine. Five o'clock, January 12, 1943. It is time to remove Tesla's brain for study in the future. You come forward, gaze into his emaciated face, and begin your work.

The scalpel feels as cold as an icicle.

"Ah," you sigh, eagerly gazing at cranial nerves 3, 4, and 5. They had orchestrated his diaphragm's contraction in livelier times. He won't be needing these you think as you pluck the nerves like banjo strings.

This is Tesla. Tesla, who harnessed the alternating electrical current used today. Telsa, who invented radio, fluorescent lighting, bladeless turbines, primitive robots. Why were his last years spent in cheap hotels feeding pigeons?

Reaching underneath the brain, you sever his spinal cord and tie a piece of twine around the base of the brain. The brain will be suspended upside down in a jar of formalin so that it will maintain its shape. In several weeks it will become more solid.

His brain is in your hand. No one will notice the missing mass when his remains are cremated at the Ferncliffe Cemetery at Ardsley-on-the-Hudson in the deep cold of a winter afternoon.

You gaze up into the sky through a broken stained-glass window and spot three pigeons, wings outstretched and motionless, flying toward the amber disk of the sun.

The sunset is like a flock of pigeons on fire.

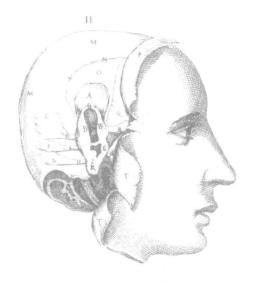

PROFILES

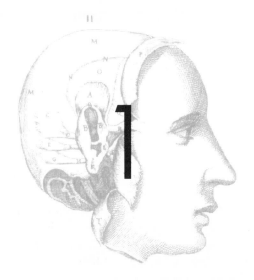

THE PIGEON MAN FROM MANHATTAN

I DO NOT THINK THERE IS ANY THRILL THAT CAN GO
THROUGH THE HUMAN HEART LIKE THAT FELT BY
THE INVENTOR AS HE SEES SOME CREATION OF THE
BRAIN UNFOLDING TO SUCCESS. . . . SUCH
EMOTIONS MAKE A MAN FORGET FOOD, SLEEP,
FRIENDS, LOVE, EVERYTHING.

NIKOLA TESLA

I DO NOT THINK YOU CAN NAME MANY GREAT
INVENTIONS THAT HAVE BEEN MADE
BY MARRIED MEN.

NIKOLA TESLA

MR. MARCONI IS A DONKEY.

NIKOLA TESLA

⬜️ FACT FILE

- Name: Nikola Tesla
- Born: 1856, Austria-Hungary (which later became Serbia). Immigrated to America in 1884 to work with Thomas Edison.
- Died: 1943, New York
- Occupation: Inventor
- Achievement: Invented the induction motor and polyphase (alternating current) power transmission, both still of considerable importance today. He was responsible for the first practical commercial use of alternating current (AC) motors, generators, and transmission lines. Tesla also developed a high-voltage generator known the "Tesla coil" which to this day is used to give spectacular lighting demonstrations. Perhaps you've seen demonstrations in museums where streamers of electricity shoot from a large metal ball.
- Marital status: Never married; celibate
- Notable physical features: Tesla was six feet, six inches tall with abnormally long thumbs. A patch of hair temporarily whitened on his jet black hair when his mother died.
- Some bizarre behaviors: Columbiphilia (pigeon-love), kakiphobia (fear of dirt), scotophilia (love of the dark), pathophobia (fear of germs), spherophobia (fear of round objects), triphilia (obsession with the number 3), and visual and auditory hallucinations
- Frivolous signature: Sometimes he signed his letters with the initials "G.I." (for Great Inventor).
- Standard dress: Black Prince Albert coat and a derby hat, white silk handkerchiefs, stiff collars—all worn even in the laboratory or when covered with living carpets of pigeons
- Residence as an adult: Various New York City hotels
- Religion: Interested in Buddhism and Christianity, but not an orthodox believer.
- Prejudice: Antisemitic, as evidenced by statements such as: "Miss! Never trust a Jew!" or ". . . though a day of plebeians—drummers, grocerymen, Jews, and other social trilobites, the prospect is nevertheless delightful."
- Despised rivals: Thomas Edison, inventor, and Guglielmo Marconi, physicist

- Favorite quotes by others about him:

> "Were we to seize and eliminate from our industrial world the results of Mr. Tesla's work, the wheels of industry would cease to turn, our electric cars and trains would stop, our towns would be dark, our mills would be dead and idle. Yes, so far reaching is his work that it has become the warp and woof of industry." (B. A. Behrend)

> "Nikola Tesla is the world's greatest inventor, not only at present but in all history. . . . His basic as well as revolutionary discoveries, for sheer audacity, have no equal in the annals of the intellectual world." (Hugo Gernsback, science editor and publisher)

- Little-known discovery: In one of Tesla's labs was a vibrating platform that Tesla discovered had a strange laxative effect. When his friend, Mark Twain, stayed on the platform too long, Tesla had to rush him to the restroom.
- Special honors: (1) The international unit for magnetic flux density is now called the Tesla, symbolized with a "T." It is equivalent to 10,000 Gauss or 1 Weber per square meter. (2) Today there is a very popular heavy-metal rock group called "TESLA." Their CD album cover has a photo of Tesla.

☞ THE STRAIGHT DOPE

> TESLA WENT BEYOND THE BORDERS OF HIS EXACT SCIENCE TO FORETELL WHAT LIES IN THE FUTURE . . . A MODERN PROMETHEUS WHO DARED REACH FOR THE STARS.
>
> LAMBERT VON BINDER

Although few people today recognize the name of Nikola Tesla, his name should be as important as Thomas Edison's in the annals of electrical technology. Before describing his strange mind and bizarre compulsions, a quick overview of his life and scientific achievements is in order.

The year 1856 was a banner one for explorers of mental and physical realms. It was the birth year not only for Nikola Tesla, but also for Sigmund

2. Nikola Tesla (1856–1943). This illustration appeared in the July 22, 1894, New York *Sunday World*, "showing the inventor in the effulgent glory of myriad tongues of electric flame after he has saturated himself with electricity."

Freud, the Austrian founder of psychoanalysis, and Robert Peary, the American explorer who discovered the North Pole. It was also the year that Big Ben, the 13.5 ton bell at the British House of Parliament, was cast, and one year before American civil engineer E. G. Otis installed the first safety elevator.

Nikola Tesla was born at midnight between July 9 and 10, 1856, in a Serbian mountain village. Tesla's father was a clergyman of the Serbian Orthodox Church. His mother could not read but was a skillful inventor of all kinds of farm implements. Tesla's parents planned for him to join the clergy, but from an early age, he was more interested in science and math. His biographers credit him with having invented a waterwheel when he was five years old.

When he got a little older, Tesla enjoyed telling scientific authorities that they were wrong and that he could do better. And most of the time he was right. For example, while a student at the polytechnical college of Graz, Austria, Tesla came upon a dynamo motor that was sparking badly between its brushes and commutator (the device used for reversing the direction of an electric current). He correctly suggested to his professor that a motor without a commutator might be created, but his professor said this was nonsense.

In 1879, Tesla enrolled at the University of Prague, but left when his father died. In 1881 he worked in Budapest for a new telephone company. During this year, he conceived of a motor that used rotating magnetic fields, paving the way for all modern polyphase induction motors. "The idea [for the motor]," Tesla said, "came like a lightening flash. In an instant I saw it all, and drew with a stick on the sand the diagrams which were illustrated in my fundamental patents of May 1888."

In 1882 Tesla went to Paris where he worked as an engineer in the Continental Edison Company, the French affiliate for Thomas Edison, the famous American inventor. The company soon sent him to Strasbourg to improve the functioning of a dynamo at a railroad station lighting plant, and at this time he built a working model of his new motor, experiencing "the supreme satisfaction of seeing for the first time rotation effected by alternating currents without commutator."

In 1884, Tesla traveled to the U.S. to promote his alternating current (AC) motor. As background, electricity can be distributed in two general ways: direct current (DC) and alternating current (AC). Direct current is a flow of current in one direction at a constant rate. On the other hand, alternating current fluctuates like a meandering river. When graphically represented, alternating current increases in magnitude from zero to a maximum, decreases back to zero, increases to a maximum in the opposite direction, returns to zero, and then repeats this process periodically. The number of repetitions of the cycle occurring each second is defined as the frequency, which is expressed in Hertz (Hz). Thomas Edison used DC exclusively and had a monopoly on many electrical patents. Tesla, however, realized that AC could be transmitted for much greater distances and knew that his AC transmission ideas would not infringe on Edison's patents. When Tesla arrived in New York, he had only four cents in his pocket and a book of poetry. As a result of a fine letter of recommendation, he was soon hired directly by Thomas Edison to redesign DC dynamos for the Edison Machine Works. Edison made it clear that he was adamantly committed to DC (not AC) power and scoffed at Tesla's ideas.

Edison and Tesla made a remarkable pair—on the exterior as different as "The Odd Couple." Tesla was impeccably dressed and feared germs. Of Edison, Tesla observed, "He had no hobby, cared for no sport or amusement of any kind, and lived in utter disregard of the most elementary rules of hygiene. If he had not married later a woman of exceptional intelligence, who made it the only object of her life to preserve him, he would have died many years ago from consequences of sheer neglect." Tesla himself didn't appear to have many hobbies, although at one point in his life he was professionally skillful at billiards.

Tesla slaved for Edison for about a year when they finally had an argument forcing Tesla to leave Edison like a bat out of hell. It all began when Tesla suggested that he could increase the efficiency of Edison's dynamos and save Edison considerable money. Edison loved the idea and replied that he would reward Tesla. "There's $50,000 in it for you," Edison said, "if you can do it." Tesla worked frantically for months, hardly sleeping. When Tesla made significant improvements, redesigned the dynamos, and installed automatic controls, he asked Edison for his $50,000 reward. Edison replied, "Tesla, you don't understand our American humor."[1]

In 1885, Tesla left Edison and began developing and selling industrial arc lamps. By 1888, Tesla had applied for many patents on various kinds of AC dynamos, transformers, and motors, and these caught the perceptive eye of wealthy industrialist George Westinghouse who bought the rights to these patents as fast as a hungry shark devouring tasty fish. Figures 3–6 contain diagrams from some of these patents. Although Edison continued to rave about the merits of direct current power transmission, Tesla's alternating current transmission methods finally triumphed when he assisted with the first large-scale harnessing of Niagara Falls. Today the world uses alternating current power, and the electric outlet in your home provides alternating current electricity.

In 1891, Tesla proudly became an American citizen. He told his friends that his certificate of naturalization was more important to him than any honorary degree. During the next few years, he worked in his New York laboratories on an array of devices. He invented the Tesla coil, which uses a spark gap capable of producing intense high-frequency discharges, and an air-core transformer. His lectures and demonstrations on high-frequency currents soon made him an international celebrity.

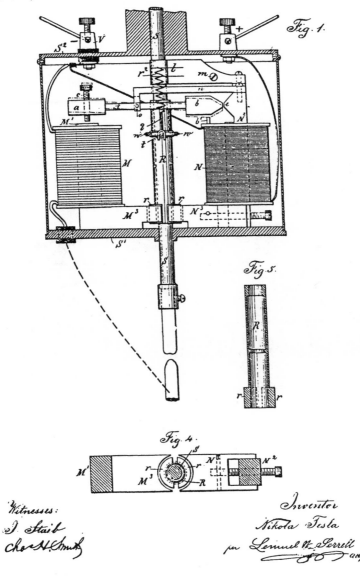

(No Model.)

2 Sheets—Sheet 1.

N. TESLA.

ELECTRIC ARC LAMP.

No. 335,786.

Patented Feb. 9, 1886.

3. Tesla's first patent was for an electric arc lamp. He applied for this patent in 1885 and received it in 1886.

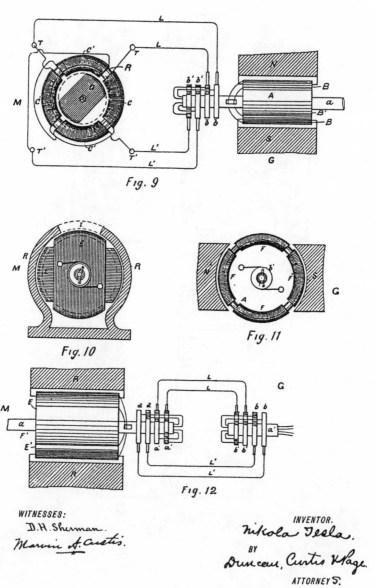

(No Model.)

4 Sheets—Sheet 2.

N. TESLA.

ELECTRICAL TRANSMISSION OF POWER.

No. 382,280.

Patented May 1, 1888.

Fig. 9

Fig. 10

Fig. 11

Fig. 12

WITNESSES:
D.H. Sherman.
Marvin A. Curtis.

INVENTOR.
Nikola Tesla.
BY
Duncan, Curtis & Page.
ATTORNEY S.

4. A diagram from Tesla's patent for the electrical transmission of power (1888).

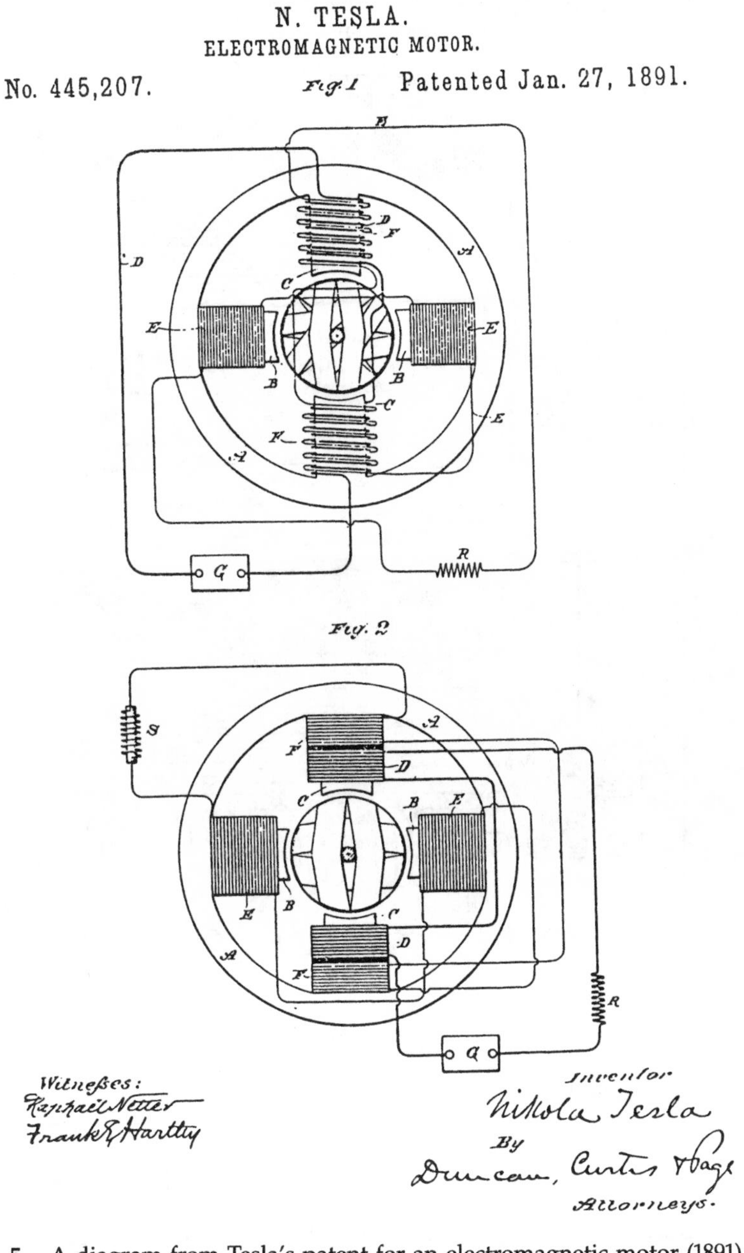

N. TESLA.
ELECTROMAGNETIC MOTOR.

No. 445,207.   *Fig. 1*   Patented Jan. 27, 1891.

Fig. 2

Witnesses:
Raphael Netter
Frank E. Hartley

Inventor
Nikola Tesla
By
Duncan, Curtis & Page
Attorneys.

5. A diagram from Tesla's patent for an electromagnetic motor (1891).

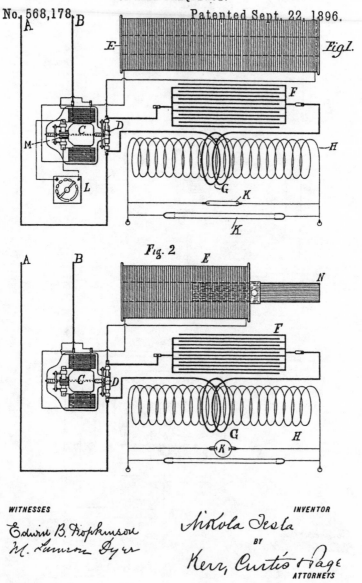

(No Model.)　　　　　　　　　　　　　　　2 Sheets—Sheet 1.

N. TESLA.
METHOD OF REGULATING APPARATUS FOR PRODUCING CURRENTS
OF HIGH FREQUENCY.

No. 568,178.　　　　　　　　　　　　　Patented Sept. 22, 1896.

WITNESSES

Edwin B. Hopkinson
M. Lawson Dyer

INVENTOR
Nikola Tesla
BY
Kerr, Curtis & Page
ATTORNEYS

6. A diagram from Tesla's patent on a method for regulating high-frequency currents (1896).

In 1893, Tesla described wireless communication devices that worked by conduction of electricity through natural media such as the Earth. In order to test his ideas, he set up a lab in Colorado Springs, Colorado, where he proved the Earth was a conductor. His method of transmitting electrical power without wires is somewhat different from radio, which uses very little energy to transmit information. For example, Tesla's dream was to pump tremendous amounts of electrical energy into the ground or air so that people could run appliances without plugging them in.

While at his Colorado laboratory, Tesla became more eccentric. At about this time he announced to the world that he had received messages from Mars using a radio receiver, which I'll describe in more detail in the "Strange Brain" section.

In additional experiments around 1899, Tesla produced artificial lighting 135 feet in length, a feat that has never been reproduced. Around 1900, Tesla started to create a worldwide communication system. With funds from wealthy industrialist J. P. Morgan, Tesla built a 200-foot transmission tower at Shoreham, Long Island. Costs escalated, and, in 1905, Morgan withdrew his financial support. The tower was later destroyed.

Although internationally famous and a millionaire in the 1890s, Tesla spent a lot of money on expensive experiments. He never bothered to seek patents for many of his inventions. During the last decades of his life, he often lived close to poverty. His final years were spent in a series of New York hotels with only pigeons for his friends.

He lived for 86 years.

STRANGE BRAIN

STANDING ALONE, TESLA PLUNGED INTO THE UNKNOWN. HE WAS AN ARCH CONSPIRATOR AGAINST THE ESTABLISHED ORDER OF THINGS.

KENNETH M. SWEZEY, SCIENCE WRITER, 1931

Overview

Let's discuss some of the more bizarre aspects of Tesla's life. For more standard treatments of Tesla's life, there are a number of fine biographies, one of my favorites being Margaret Cheney's *Tesla: Man Out of Time*. There are also some biographies as unusual as Tesla himself, because Tesla's curious behavior and inventions were magnets for an assortment of odd and inquisitive aficionados. For instance, Margaret Storm's 1959 biography of Tesla is printed in green ink and asserts that he was a superior being from the planet Venus. I list a number of other books and papers in the "Further Reading" chapter at the end of this book. Since much of the information on Tesla's life comes from Tesla himself and since he promoted himself as the "great inventor," I urge to you to view some of his more boastful claims with skepticism.

First, I'll give you a brief sampling of some of his compulsions. Throughout his adult life, Tesla lived in numerous hotels and was always elegantly attired. Picture in your mind his splendidly clothed countenance as he chews on his specially prepared meals in the Palm Room of the Waldorf-Astoria. Watch how he stacks 18 napkins in a neat little pile because he favors numbers divisible by three. If you were to keep an eye on his maid as she cautiously entered his room, you would see her carrying precisely 18 clean towels each day—at Tesla's strict request. If asked why, Tesla can shed no light. But table accoutrements and towels are not the only items he demanded be in multiples of three. He often felt compelled to walk around the block three times. He always counted his steps while walking. He chose room number 207 in another of his residences, the Alta Vista Hotel, because 207 is divisible by 3. He wouldn't even allow the maid to dust, because he liked to do his own cleaning. He washed his hands excessively and had a morbid fear of germs. In times of need, he had the ability to go for days without sleep.

Tesla had what modern psychologists call obsessive–compulsive disorder. People afflicted with this can't resist performing certain odd rituals. For example, Tesla insisted his meals be served by the maitre d'hotel rather than a waiter, and he always telephoned ahead of time with special instructions for his meals. He couldn't enjoy his meal without first lifting each napkin and then discarding it to form a large pile on the table. After each dish arrived, he compulsively counted the number of items before eating. For example, he would look in coffee cups, soup plates, and other containers and count anything in the shape of cubes. If he tried to resist this impulse, he couldn't enjoy the meal. These are all classic signs of obsessive–compulsive disorder, thought to be caused by a brain chemistry imbalance (which I'll discuss in detail in Chapter 10).

Tesla was repulsed by pearls, although other sharp-edged jewelry and gems gave him no cause for alarm. Generally speaking, he detested earrings on women. But not only did Tesla have certain visual sensitivities and abhorrences, he also was hypersensitive to certain odors. The slightest scent of camphor would drive him wild. He also was a synesthesiac: someone with unusual sensory cross-overs. For example, during his research, if he were to drop small squares of paper in a dish filled with liquid, a horrible taste would fill his mouth. He couldn't touch other peoples' hair "except perhaps at the point of a revolver."

THE EARLY DAYS

> TESLA BELIEVED THAT THE MOST ENDURING WORKS
> OF ACHIEVEMENT HAVE COME FROM
> CHILDLESS MEN.
>
> KENNETH M. SWEZEY, SCIENCE WRITER, 1931

Soaring

How did Tesla's strange behaviors begin, and what impact did they have on his creativity? In studying Tesla's life, the first thing you'll notice is that he was quite inventive, even as a young boy. His least successful youthful experiment involved human flight by umbrella. On one beautiful breezy day the child Tesla precariously perched himself on the roof of a barn. While tightly grasping his mother's large umbrella, he intentionally took rapid, deep breaths until he felt light and dizzy. Next he leaped from the roof. I don't think anyone knows how long he lay there unconscious on the moist ground, but eventually his mother picked him up and carried him home.

The desire to fly has appealed to humankind since the dawn of civilization, and Tesla's strong desire to fly persisted throughout his life. For example, when he was eleven he continuously dreamed of flying machines: "Every day," he said, "I used to transport myself through the air to distant regions but could not understand just how I managed to do it."

Slowly his dreams became more concrete. He had visions of flying machines made of rotating shafts, flapping wings, and powerful vacuums. "From that day on," Tesla writes of his dreams, "I made my aerial excursions in a vehicle of comfort and luxury as might have befitted King Solomon."

Tesla's visions of flying continued into adulthood, and finally, two days after New Year's Day in 1928, Tesla obtained a patent for his vertical takeoff aircraft, diagrammed in Figure 7. This patent was one of the last patents of his life.

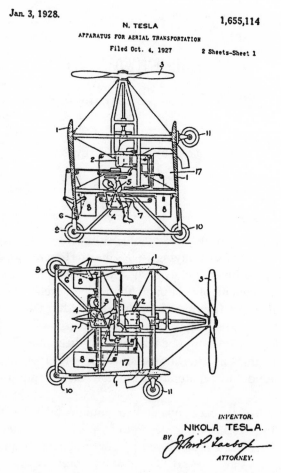

7. One of Tesla's last patents: a vertical takeoff aircraft, 1928.

Tesla's dangerous, youthful experiments weren't confined to flight. In fact, he seemed to have been involved in an inordinate number of risky situations. For example, he writes that he was once almost cremated and boiled alive in milk. He was trapped in tombs and attacked by packs of large crows and sharp-tusked hogs.

I admire Tesla's adventuresome spirit, although his umbrella experiments were about as successful as his insect-powered motors. Tesla would spend hours gluing June bugs to small windmill contraptions. The trapped creatures would beat their wings in order to power the device. Unfortunately, the experiments abruptly ended when a young friend of Tesla's displayed a peculiar hunger for June bugs. The child shoved them into his mouth like clumps of little raisins. They made a sick crunching sound. After carefully watching the boy chewing on the wriggling beetles, Tesla promptly vomited, and the June bug windmill was put away.

Visions

After his brother died, Tesla pursued his schooling with iron discipline bordering on madness. His goal was to outdo his fellow students. Could the ferocious demands he placed on himself have exacerbated his eccentricities? Like Isaac Newton, who had suddenly decided to approach his schooling with a maniacal drive to be superior to other students, Tesla was single-minded and had extraordinary powers of concentration. When Tesla decided to read the complete works of Voltaire, he initially didn't realize that there were 100 volumes printed in a small font. Once he started, he could not rest until he completed his goal. The task nearly drove him to a mental breakdown.

In school, Tesla startled the teachers with his hyperspeedy answers. A teacher would begin to write a mathematical problem on the board, and just as she was finishing, Tesla would write the answer. Although teachers first suspected him of cheating, they soon realized that his intense concentration and photographic memory of logarithmic tables allowed him to solve mathematical problems more quickly than ordinary students.

Since childhood, Tesla and his brother saw visions and bright lights.[2] Tesla wrote, "this peculiar affliction of the appearance of images is often accompanied by strong flashes of light which marred the sight of real objects and interfered with my thoughts. They were pictures of things and scenes which I had really seen, never of those I imagined. When a word was spoken to me the image of the object it designated would present itself vividly to my vision, and sometimes I was quite unable to distinguish whether what I saw was tangible or not. This caused me great discomfort and anxiety." Tesla later speculated about inventions that could make prints of his thoughts. "It should be possible," Tesla said, "to project on a screen the image of any object one conceives and make it visible."

As a boy, Tesla claimed to have additional visual sensitivities, and most intriguing was his hypersensitivity to snow. "People walking in the snow," he said, "left a luminous trail behind them, and a snowball thrown against an obstacle gave a flare of light like a loaf of sugar hit with knife."

One day while petting his cat named Macak, Tesla saw something even more disconcerting than luminous snow. "Macak's back was a sheet of light, and my hand produced a shower of crackling sparks loud enough to be heard all over the place." Tesla's father explained to him that the sparks were caused by electricity.

Tesla began to wonder about this mysterious phenomenon. "Is nature a gigantic cat?" he writes. "If so who strokes its back? It can only be God." Later that night, Tesla perceived an aura around the cat.

In His Mind's Eye

To escape his tormenting dreams and visual weirdness, Tesla would spend hours creating imaginary mental worlds. In these parallel universes, he would make new friends, go on wonderful journeys. His alternate realities with their villages and people were so realistic that he wrote, "however unbelievable, it is a fact that they were just as dear to me as those in actual life and not a bit less intense in their manifestations."

When it came to the construction of sophisticated devices, he often performed a "virtual reality" simulation in his mind's eye without ever touching a workbench. He mentally designed machines by simply picturing all the various parts of a new device moving and coming together. "I do not rush into actual work," he said. "When I get an idea I start at once building it up in my imagination. I change the construction, make improvements and operate the device in my mind. It is absolutely immaterial to me whether I run my turbine in my thought or test it in my shop. I even note if it is out of balance."

While Tesla's method of visual invention helped him immensely, it had one drawback. Often after mentally creating his inventions, he would never relegate them to a working model or drawing needed for commercial success. If Tesla's brain were available to us to explore his ancient ridges like archeologists, what inventions would we mine and uncover?

When Tesla was twelve years old, he was able to suppress the bothersome images flashing across his mind but he was never able to completely control the flashes of light and flame-like protuberances that occurred in times of great emotion. At the age of sixty, Tesla reported:

> These luminous phenomena still manifest themselves from time to time, as when a new idea opening up possibilities strikes me, but they are no longer exciting, being of relatively small intensity. When I close my eyes, I invariably observe first a background of very dark and uniform blue, not unlike the sky on a clear but starless night. In a few seconds this field becomes animate with innumerable scintillating flakes of green, arranged in several layers and advancing towards me. Then there appears, to the right, a beautiful pattern of two systems of parallel and closely spaced lines, at right angles to one another, in all sorts of colors with yellow-green and gold predominating. Immediately thereafter the lines grow brighter and the whole is thickly sprinkled with dots of twinkling light. The picture moves slowly across the field of vision and in about ten seconds vanishes to the left, leaving behind a ground of rather unpleasant and inert grey which quickly gives way to a billowy sea of clouds, seemingly trying to mold themselves in living shapes. It is curious that I cannot project a form into this grey until the second phase is reached. Every time, before falling asleep, images of persons or objects fly before my view. When I see them I know that I am about to lose consciousness. If they are absent and refuse to come it means a sleepless night.

Insights

Many creative people act on hunches and insight rather than analytic assessment. For example, R. H. Davis writes on Einstein's intuition in a 1995 issue of *The Skeptical Inquirer*:

> Einstein's fundamental insights of space/matter relations came out of philosophical musings about the nature of the universe, not from rational analysis of observational data—the logical analysis, prediction, and testing coming only after the formation of the creative hypotheses.

Tesla also did not conceive of his inventions using a logical system. Sometimes, while not fully concentrating on a problem, he would "know" that a solution would soon arrive. He writes, "and the wonderful thing is that if I do feel this way, then I know I have really solved the problem and shall get what I am after."

As a young man he conceived of many unusual inventions that sound impractical today. One of his most startling plans was a mail tube stretching across the Atlantic Ocean. To use it, one would insert the letter in a spherical water-tight container. A huge pump would force the water through the tube, thereby shooting the letter from the U.S. to Europe. Unfortunately, the device could never be made because the friction of water flowing through such a long pipe would render the concept unworkable. (We will encounter a similar idea of transporting objects with pipes in Chapter 8 where I discuss inventor Geoffrey Pyke, who, in the 1940s, suggested a pipe method for conveying soldiers from ships to shore.)

Tesla looked at many facets of life in terms of inventions. Humans were automatons, "meat machines" responding like robots to their environment. People had no wills. Their acts were simply mechanical reactions to external stimuli, like the liquid in a thermometer responding to heat. Later in life, Tesla wrote, "We are but cogwheels in the medium of the universe, and it is an unavoidable consequence of the laws governing that the pioneer who is far in advance of his age is not understood and must suffer pain and disappointment and be content with the higher reward which is accorded to him by posterity."

Of Flies and Pigeons

In 1881, Tesla's hearing became unusually acute—or at least this is what he tells us. While some of this hypersensitivity is plausible, such as being able to hear a ticking clock three rooms away, it is astonishing that many Tesla biographers fail to challenge his more extreme assertions. The only diseases of which I am aware that could cause such hypersensitivity are hyperacusis or Williams syndrome. People with Williams syndrome have abnormally acute hearing, often resulting in pain on hearing only moderately loud sounds. Many of the biographies of Tesla might be called hagiographies because extraordinary claims made by Tesla go unchallenged. For example, at one point in his life, Tesla said he perceived the landing of a fly on a nearby table as a loud thud. He said train whistles 20 miles away made his chair vibrate, and he often had to put rubber casters under the legs of his bed to damp unwanted vibrations. He wrote:

> The roaring noises from near and far often produced the effect of spoken words which would have frightened me had I not been able to resolve them into their accidental components. The sun's rays, when periodically intercepted, would cause blows of such force on my brain that they would stun me. I had to summon all my willpower to pass under a bridge or other structure as I experienced a crushing pressure on the skull. In the dark I had the sense of a bat and could detect the presence of an object at a distance of 12 feet by a peculiar creepy sensation on the forehead.

Some time during 1881 his pulse was said to fluctuate wildly, starting very low and then going as high as 260 beats per minute. The continuous twitching of his skin became torture for him.

Before concluding this section, note that Tesla's earliest years were spent at his family's Serbian countryside home, which was always surrounded by birds. Pigeons continually cooed in the cote. Chickens clucked. Tesla would spend hours watching the precise formations of geese in the sky. In my opinion, his fascination and incredible love for pigeons at the end of his life was a subconscious yearning for fond childhood memories. As an old man he would warm nearly frozen pigeons in his hands and tell his friends, "All things from childhood are still dear to me."

LIFE IN AMERICA

> I LIKED TO FEED OUR PIGEONS, CHICKENS, AND OTHER FOWL, TAKE ONE OR THE OTHER UNDER MY ARM AND HUG AND PET IT.
>
> NIKOLA TESLA

From Street Gang to Stardom

In 1884, Tesla came to America and worked for Thomas Edison, but as previously described soon resigned over a financial dispute. Tesla's engineering reputation grew, and he soon formed the Tesla Electric Light Company in Rahway, New Jersey. Here he developed the Tesla arc lamp, which was safer and more economical than those currently in use. Unfortunately, Tesla was slowly eased out of the company by his investors and was unable to find any engineering job. Due to the depression, his stock certificates had little value. In 1887, Tesla toiled as a laborer in New Street gangs and barely managed to survive. He seldom referred to his painful experience, so we do not have details of his life as a laborer.

A manager of the Western Union Telegraph Company grew interested in Tesla's ideas involving AC. With the manager's help, Tesla set up a new company, The Tesla Electric Company, with the goal of developing practical AC applications.

In 1891, Tesla applied for 40 U.S. patents dealing with AC motors and the like. Tesla was finally recognized for his brilliant ideas, and George Westinghouse, the Pittsburgh magnate, soon purchased Tesla's patents for about $60,000. Tesla began to consult for Westinghouse for $2,000 per month.

As noted earlier, Thomas Edison believed that DC applications were the wave of the future, and he did all he could to persuade the public of the dangers of AC. He actually paid school boys 25 cents for every pet they could steal from neighbors of his West Orange, New Jersey, plant. Edison had the animals electrocuted and displayed in order to scare away potential users of AC.[3] So intent were Edison and his team on discrediting the use of AC power that one of his assistants went on a travelling road show electrocuting calves and small dogs. Their nervous moos and barks were one-by-one abruptly terminated. Edison said he was "Westinghousing" the animals.

New York State scheduled the first "humane" electrocution of a condemned murderer on August 6, 1890. However, due to miscalculations by Edison's engineers, the electric current was too weak, and the condemned man was only "half-killed." Over a period of 10 to 15 minutes the man was fried and refried in a device similar to the one in Figure 8.

Amnesia in the 1890s

Tesla began to develop a strange amnesia in the 1890s, perhaps due to intense stress as a Westinghouse consultant or his manic desire to create a radio that worked by transmitting signals through the Earth. He no longer

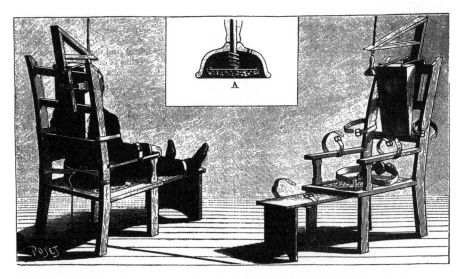

8. Front and rear view of an electric chair used during Tesla's time. Inset A is the electrode that is pressed against the head (1890).

could visualize past events except those from infancy. He wrote, "I had produced a striking phenomenon with my grounded transmitter and was endeavoring to ascertain its true significance in relation to the currents propagated through the Earth. It seemed a hopeless undertaking, and for more than a year I worked unremittingly but in vain. This profound study so entirely absorbed me that I became forgetful of everything else, even of my undermined health. At last, as I was on the point of breaking down, nature applied the preservative, inducing lethal sleep." Tesla began to sleep "as if drugged." He could no longer recall any scenes from the last 30 years.

Tesla hated to go to doctors and therefore tried to cure his amnesia using his own methods. Each night he concentrated on memories, trying to bring them into focus. During his memory recovery progress, he seemed to become more and more attached to his mother. He wanted to visit her. "The feeling grew so strong," Tesla recalled, "that I resolved to drop all work and satisfy my longing. But I found it too hard to break away from the laboratory, and several months elapsed, during which I succeeded in reviving all the impressions of my past life."

Even during his amnesiac phase, Tesla still had a photographic memory for his current technical work. "I could recall the smallest details," he said, "and the least significant observations in my experiments, and even recite pages of text and complex mathematical formulae."

In February 1892, he received a telegraph that his mother was dying. He rushed to Europe just in time to spend a few hours with his mother. During the night, he had a vision of "a cloud carrying angelic figures of marvelous beauty, one of whom gazed upon me lovingly and gradually assumed the features of my mother. The appearance slowly floated across the room and vanished, and I was awakened by an indescribably sweet song of many voices. In that instant a certitude, which no words can express, came upon me that my mother had just died. And that was true . . ."

ESP, Precognition, and the Occult

Tesla believed in extrasensory perception and felt he could perceive the pain of others. Around the year 1892, he wrote: "A very sensitive and observant being, with his highly developed mechanism all intact, and acting with precision in obedience to the changing conditions of the environment is endowed with a transcending mechanical sense, enabling him to evade perils too subtle to be directly perceived. When he comes in contact with others whose controlling organs are radically faulty, that sense asserts itself and he feels the cosmic pain."

At the Chicago World's Fair of 1893, Tesla applied 200,000 volts to his body, and reporters said that his clothing and body continued to glow after the power was cut off. Tesla claimed that such currents could keep a naked man warm at the North Pole. He also believed that electrical anesthesia would some day be possible, and he suggested that bored schoolchildren be stimulated by wires running along the floors. In order to pep up actors before a performance, he suggested that their dressing rooms be electrified.

Occultists were drawn to Tesla like groupies to a rock star. An assortment of odd individuals tried to associate with Tesla, and some claimed he was not of terrestrial origin but rather a Venusian who traveled to Earth either on a spaceship or the wings of a white dove. His mission, they said, was to advance the human race. Tesla did not welcome this strange form of adoration and disavowed the special powers they attributed to him.

Tesla's wonderful electrical demonstrations with glowing tubes and electrical discharges reached a feverish pitch in the 1890s, and some of my favorite illustrations of his public experiments are shown in Figures 9–11.

9. Tesla, the archangel. One of Tesla's most striking demonstrations involved 3-foot-long sealed glass tubes known as Geissler tubes. When Tesla moved the tubes into the field between 10-foot charged bars on the ceiling and floor, the tubes became luminous. In the words of one reporter, "Tesla stood there, like the archangel, brandishing the flaming sword!" (1893).

Some Fears

In 1893, Tesla began to dine regularly at the Waldorf–Astoria hotel. Sometimes he attended sumptuous dinners where people smoked tobacco wrapped in $100 bills and hid jewels in the folds of napkins to surprise the ladies. Tesla was always impeccably dressed. Even in his laboratory, he wore a black Derby hat and used silk handkerchiefs. His collars were stiff. He threw out gloves after wearing them a few times, probably because of his fear of germs. He never wore jewelry, being afraid of most forms of jewelry.

Tesla compulsively loathed fat people. When his overweight secretary knocked something on the floor, he fired her. When she got on her knees and begged for a second chance, he declined.

Tesla's experiments [1894]

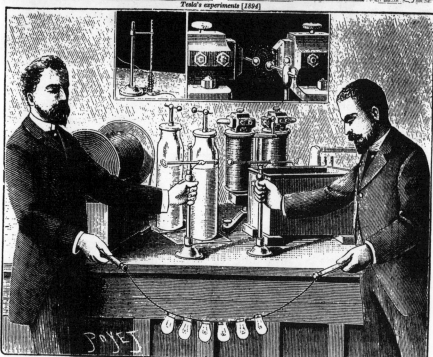

10. Tesla experiments.

Recent Tesla experiments in Berlin [1894]

11. Tesla experiments in Berlin (1894).

Perpetual, Terrestrial Night Light

One of Tesla's greatest dreams was to eliminate the night. By introducing high-frequency currents into the upper atmosphere, he hoped to light up entire cities, countries, the entire world. He hypothesized that the atmosphere could be made to glow like gas in a glass tube. Airplanes would never have to land in the dark, and shipping lanes would be safer. Tesla never revealed the mechanism by which he would conduct current to the upper air. Journalists speculated that he planned to shine ultraviolet light into the atmosphere, ionizing the air and making it a good conductor of electricity. This conducting path might be used to transmit his high-frequency currents.

Tesla's Favorite Invention

In 1893, Tesla fell in love with a simple invention that you can buy today for less than $5 from a novelty shop. The device was a small heat-powered windmill called a radiometer. When sunlight falls upon the windmill's vanes inside a bulb, the windmill begins to spin. Tesla said he considered the device "the most beautiful invention made."

For those of you who are interested in the physics and history of radiometers, they were invented by Sir William Crookes in the late 1800s. To create a radiometer, evacuate most of the air from a glass bulb creating a partial vacuum. Mount the rotor on a vertical support inside the bulb. The rotor has four light, horizontal arms attached at right angles to one another on a central pivot. Each arm has a vane that has a black surface and a white surface. When light strikes on the white surface, most is reflected away. However, the black surface absorbs light and heats up. Nearby air molecules are heated and exert a pressure on the black surface, causing the rotor to turn.

X-Rays

Tesla was a wild man when it came to x-rays. Around 1896, he began experimenting with these invisible rays that were discovered in 1895 by Roentgen. Tesla claimed that he was making 40 minute exposures of the human head from an x-ray device 40 feet away. If this was true, it would be quite an accomplishment for that time. Tesla did not appear to be aware of the dangers of x-rays, and repeatedly exposed his head to x-rays to "stimulate" his brain. He writes, "An outline of the skull is easily obtained with an exposure of 20 to 40 minutes. In one instance an exposure of 40 minutes gave clearly not only the outline, but the cavity of the eye, the lower jaw and connections to the upper one, the vertebral column and connections to the skull, the flesh, and even the hair."

When his skin began to resemble something from *Night of the Living Dead*, Tesla realized that x-rays were not so innocuous. "In a severe case," he wrote, "the skin gets deeply colored and blackened in places, and ugly, ill-foreboding blisters form; thick layers come off, exposing the raw flesh. . . . Burning pain, feverishness, and such symptoms are natural accompaniments. One single injury of this kind in the abdominal region of a dear and zealous assistant—the only accident that ever happened to anyone but myself in all my laboratory experience—I had the misfortune to witness."

More on Women's Pearls

In 1893, Tesla was introduced to the beautiful Anne Morgan, daughter of the wealthy industrialist J. P. Morgan. Tesla took one look at her and became so disgusted by her pearl earrings that he began to grind his teeth together and did his best to avoid her. He would have enjoyed talking with her, but the pearls in particular made this impossible. As might be expected, Tesla's friends always asked him why he didn't marry. Not only were his friends worried about his disinterest in women, but so was an entire nation. Unbelievably, various technical journals even took up the rallying cry to get Tesla married. These journals included: the *Electrical Journal*, the *Electrical Review of London*, and the *American Electrician*.

Like many inventive minds—for example, Newton, Michelangelo, Sir Francis Bacon, and Oliver Heaviside—Tesla never married. When a reporter asked Tesla if he believed in marriage, he responded, "For an artist, yes; for a musician, yes; for a writer, yes; but for an inventor, no. The first three must gain inspiration from a woman's influence and be led by their love to finer achievement, but an inventor has so intense a nature with so much in it of wild, passionate quality, that in giving himself to a woman he might love, he would give everything from his chosen field. I do not think you can name many great inventions that have been made by married men."

Heart Ache

Tesla was often depressed because he had no time to develop all the ideas swimming around in the ocean of his mind. He writes in the third person about himself, "So many ideas go chasing through his brain that he can only seize a few of them as they fly, and of these he can only find the time and strength to bring a few to perfection. And it happens many times that an inventor who has conceived the same ideas anticipates him in carrying one out of them. Ah, I tell you, that makes a fellow's heart ache."

Tesla had an extreme emotional low in 1895 when his lab burned down. At the time, he was very close to finding a way to economically store and produce liquid oxygen. Because of Tesla's setback Carl Linde, a German scientist, beat him to the discovery. "I was so blue and discouraged in those days," Tesla said, "that I don't believe I could have borne up but for the regular electric treatment which I administered to myself. You see, electricity puts into the tired body just what it most needs—life force, nerve force. It's a great doctor, I can tell you, perhaps the greatest of all doctors."

Life on Mars

Mars, the first planet in the solar system beyond Earth's orbit, is a comparatively small world. Before the advent of the Space Age, it was considered to be a planetary twin of Earth with comparable temperatures, and containing water, at least in the form of ice at its poles. The enigmatic red planet has fascinated humans since the ancient Greeks. Galileo made the first telescopic observations of Mars in 1610. Four years after Tesla was born, E. Liais, a French astronomer living in Brazil, announced that dark patches on Mars were old seabeds filled with primitive vegetation. In 1877, G. V. Schiaparelli in Milan drew a detailed map of Mars showing a number of artificial-looking lines he called canals. He suggested that they were part of a global irrigation system designed to transport water from the polar caps to warmer regions. Today we know that such a canal system does not exist and that the earlier suggestions were due to telescopic aberration or wishful thinking.

Tesla often gazed up into the sky and speculated on the possibility of Martian life-forms. He believed that intelligent beings existed on Mars. Moreover, he wanted to communicate with them using a radiolike device

N. TESLA.
APPARATUS FOR TRANSMITTING ELECTRICAL ENERGY.
APPLICATION FILED JAN. 18, 1902. RENEWED MAY 4, 1907.

1,119,732.

Patented Dec. 1, 1914.

12. A diagram from Tesla's patent for a device that transmits electrical energy through the air (1914).

like the one shown in Figure 12 for "the transmission of power from station to station without the employment of any connecting wire." He had proved to himself that electrical energy could be transmitted through the upper atmosphere at great distances. He realized that very large voltages caused chemical reactions involving atmospheric oxygen and nitrogen, and he worried about igniting the atmosphere. "So energetic are these actions," he said, "and so strangely do such powerful discharges behave that I have often experienced a fear that the atmosphere might be ignited, a terrible possibility. Who knows but such a calamity is possible?"

Earth Cracker

Tesla claimed he had the awesome power to crack the Earth like a nut in a nutcracker. It all began in 1898, when Tesla became interested in mechanical vibrations and created several pocket-sized oscillators. Margaret Cheney in her book *Tesla: Man Out of Time* tells a story of Tesla placing one of these devices on a building's iron pillar. In less than minute, the oscillator began to resonate with the pillar causing huge vibrations. A small earthquake started and Tesla had to smash the oscillator with a hammer to save the building. He said to police who had just arrived on the scene, "Gentlemen, I am sorry. You are just a trifle too late to witness my experiment. I found it necessary to stop it suddenly. . . . However if you will come around this evening I will have another oscillator attached to this platform and each of you can stand on it. You will, I am sure, find it a most interesting and pleasurable experience. Now you must leave, for I have many things to do. Good day, gentlemen."

When reporters arrived, Tesla boasted he could easily destroy the Brooklyn Bridge with one of his small vibrators if he had the notion to do so. Later he told friends that he could split the entire Earth using similar devices—"split it as a boy would split an apple—and forever end the career of man." Tesla said:

> If I strike the earth at this instant, a wave of contraction goes through it that will come back in one hour and forty-nine minutes in the form of expansion. As a matter of fact, the earth, like everything else is in a constant state of vibration. It is constantly contracting and expanding. Now, suppose that at the precise moment when it begins to contract, I explode a ton

of dynamite. That accelerates the contraction and, in one hour and forty-nine minutes, there comes an equally accelerated wave of expansion. When the wave of expansion ebbs, suppose I explode another ton of dynamite, thus further increasing the wave of contraction. Is there any doubt as to what would happen? There is no doubt in my mind. The earth would be split in two. For the first time in man's history, he has the knowledge with which he may interfere with cosmic processes!

When asked how much time it would take to split the Earth into two pieces, he said, "Months might be required; perhaps a year or two." Tesla called this new science of earth vibration "telegeodynamics."

Tesla tried to commercialize a communication machine that would transmit messages through the Earth by banging the ground with a steel cylinder. Such messages, he said, would be unaffected by the weather and atmospheric disturbances. The mechanical waves produced by the device were supposedly not attenuated much as they traveled. Tesla reasoned that people from all over the word could therefore communicate with one another by banging the ground like primitive tribes drumming.

No one developed the device, but Tesla never lost interest in the use of mechanical resonances. He bragged to reporters that he could reduce the Empire State Building to rubble using "an engine so small you could slip it in your pocket." At first, the outer stone structure would crumble, followed swiftly by the steel supports.

He believed that future wars would not be waged with guns and bombs but with electrical waves.

Messages from Mars

In 1900, Tesla received a message from Mars. At least that's what he thought and announced to the world. While working in his lab on a powerful radio receiver, he heard some strange rhythmic sounds. He reasoned that the regular patterns could only be produced by intelligent life residing on Venus or Mars. Today radio astronomers know that regular patterns are often emitted from certain types of stars such as pulsars.

British physicist Lord Kelvin (1824–1907) announced to the world that he agreed with Tesla that Mars was signaling America. He also said that New York was the "most marvelous lighted city in the world" and the only city visible to the Martians. "Mars is signalling New York," Kelvin said. Tesla's friend Hawthorn wrote that the Martians had visiting the Earth for years but chose not to communicate with ignorant humans until someone as advanced as Tesla was born. Hawthorn wrote, "Possibly [the starry men] guide his development; who can tell?"

Scientists at the time ridiculed Tesla and asked him to show them his "extraterrestrial ear." They couldn't believe such claims without seeing his receiving device in action. However, Tesla desired to keep the apparatus secret.

Dark Man

Around 1913, Tesla wanted to live in the dark. He yearned for Stygian blackness like some nocturnal creature prowling the depths of Hades. When he arrived in his office around noon, he would promptly close all the shades. Apparently he was most productive in the dark, and his employees would frequently hear him talking to himself in the dark office. Despite his scotophilia (love of the dark), Tesla's creativity and genius were unimpaired. His scotophilia seems to be in contradiction with his earlier desire to eliminate the night using a perpetual, terrestrial night light.

We may add scotophilia to Tesla's other strange pathologies: pathophobia (abnormal fear or dread of disease or germs) and kakiphobia (fear of dirt).

Religion

Most of the greatest minds through history believed in God and were interested in religion. Astronomer Fred Hoyle once noted, "I have always thought it curious that, while most scientists claim to eschew religion, it actually dominates their thoughts more than it does the clergy."

Tesla was not an orthodox believer in any one religion although he was fascinated by Buddhism and Christianity. He commented that religion was an excellent thing for *other* people. He predicted that by the year 2100 the entire Earth would have an amalgam of Buddhist and Christian beliefs.

Tesla did believe in the likelihood of life after death. He once remarked, "the greatest mysteries of our being are still to be fathomed and that . . . death itself may not be the termination of the wonderful meta-morphoses we witness."

Drinking

> THOUS HAST WYLDE LYCOURE, THE WHICHE MAKETH
> ALL THY STOMACKE TO BE ON A FLAMBE.
> B. SKELTON, 1529

In the 19th century, U.S. temperance movements lobbied for laws preventing the manufacture and distribution of alcoholic beverages. The Prohibition party made it a national issue in 1869, and during World War I, the idea of prohibition rose to prominence when conservation policies limited the production of liquor. In 1919, the 18th amendment to the Constitution established Prohibition, and it affected almost all facets of American society. Tesla himself was a drinking man and loathed the idea of Prohibition.

Tesla frequently drank small amounts of ambrosia, and when Prohibition was instituted, he thought the government was crazy to invade his privacy. To Tesla, alcohol was a healthy elixir. With a governmental ban, he believed that his health would deteriorate. Tesla was so distressed that during Prohibition he withdrew his long-standing prediction that he would live to the age of 140. (He lived to 86.)

The Pigeon Man

Sometimes when I think of Tesla I can't help but seeing in my mind the Birdman of Alcatraz. More often I think of Alfred Hitchcock's *The Birds*.

Towards the end of his life, Tesla became a columbiphilic, that is, he fell in love with pigeons. Little, big, black, white, healthy, sick, fat, thin, noisy, quiet. They all came from the New York streets, alleys, and parks. Tesla spent a good deal of his time collecting sick pigeons and nursing them back to health at the Waldorf–Astoria where he lived for years on credit because he had no money. While searching for his feathered friends, he would sometimes coax them to take food from his lips—quite a contradiction for a man usually fearful of germs. People saw him feeding pigeons individually or in huge groups. Sometimes he covered himself from head to toe with flapping creatures. Margaret Cheney writes, "They perched on his head, pecked feed from his hands, and covered his arms in a living, gurgling carpet of birds swarming over his black evening pumps." Tesla called pigeons his "sincere friends."[4]

In 1929, a science writer named Kenneth Swezey came to interview Tesla. Tesla took the writer on one of his midnight sojourns through the city. As they walked, Tesla spoke about his ideas for wireless communication but then stopped suddenly. "However, what I am anxious about at this moment is a little sick bird I left up in my room. It worries me more than all my wireless problems put together." Tesla explained that he had found a pigeon that had difficulty eating because of its crooked beak and cancerous growth on its tongue. Tesla was certain he could cure his avian ally.

Tesla also had a bird menagerie in his hotel room at the St. Regis. The birds that he could not put in his hotel room were sent to a nearby bird shop. Some had broken wings or legs. Not unexpectedly, the maids and servants complained about the birds.

In 1921, Tesla became sick in his office. Fearing that he would not be able to return to his St. Regis Hotel room, he had his secretary call the hotel's housekeeper to take care of his favorite companion, "the white pigeon with touches of gray in her wings." In the past when Tesla was unable to feed the birds behind the New York City Public Library in Bryant Park, he actually hired a Western Union messenger to do the job for him. Later when his favorite (female) white pigeon became ill, Tesla didn't stray from his hotel room until he had nursed it back to health.

All relationships finally come to an end, even interspecies liaisons. A year later when the white bird finally died, he walked into his office carrying its corpse wrapped in cloth. Tesla was close to tears when he asked a friend in the suburbs to bury the bird on his property where the grave could be properly cared for.

When his friend was about to deposit Tesla's beloved in a shallow grave, he received a call from Tesla. "Bring her back please," Tesla said. "I have made other arrangements." To this day, we do not know how Tesla disposed of the body, but it is clear he had a special, mystical, perhaps unwholesome, attachment to the bird. Tesla wrote:

> I have been feeding pigeons, thousands of them, for years. But there was one pigeon, a beautiful bird, pure white with light grey tips on its wings; that one was different. It was a female. I

would know that pigeon anywhere. No matter where I was that pigeon would find me, when I wanted her I had only to wish and call her and she would come flying to me. She understood me and I understood her. I loved that pigeon. Yes, I loved her as a man loves a woman, and she loved me. When she was ill I knew, and understood; she came to my room and I stayed beside her for days. I nursed her back to health. That pigeon was the joy of my life. If she needed me, nothing else mattered. As long as I had her, there was a purpose in my life. Then one night as I was lying in my bed in the dark, solving problems, as usual, she flew in through the open window and stood on my desk. I knew she wanted me; she wanted to tell me something important so I got up and went to her. As I looked at her I knew she wanted to tell me—she was dying. And then, as I got her message, there came a light from her eyes—powerful beams of light. . . . Yes, it was a real light, a powerful, dazzling, blinding light, a light more intense than I had ever produced by the most powerful lamps in my laboratory. When that pigeon died, something went out of my life. Up to that time I knew with a certainty that I would complete my work, no matter how ambitious my program, but when that something went out of my life I knew my life's work was finished.

Some psychologists have analyzed Tesla's columbiphilia. The bird is said to symbolize Tesla's mother, whom he dearly missed. Dr. Jule Eisenbud wrote in the July 1963 issue of the *Journal of the American Society for Psychical Research* that the dove represented Tesla's mother's nourishing breast. Tesla's mother did in fact report that Tesla went into a deathlike depression when she took him off her breast at the age of two weeks. Eisenbud concludes that Tesla could not tolerate smooth round surfaces such as pearls because they reminded him of breasts. He detested all smooth round objects. Later in life he could not bring himself even to say the word "sphere." Toward the end of his life, he lived mostly on warm milk.

I find some paradoxical behavior in Tesla's columbiphilia. For example, why is it that he loved pigeons, but one of his favorite foods was squab? Interestingly, he was rather particular about what part of the bird he would eat. For example, he wouldn't eat the entire squab, but only the meat on either side of the breastbone.

Three years after his pigeon companion died, Tesla still lived at the St. Regis Hotel. Tesla had always been very generous with money, giving away millions over the course of his life. Somehow he thought he could always make more money through his inventions. But now, he couldn't pay his bills, and the police were sent to evict him. Tesla wanted to make things right with his secretaries whom he hadn't paid in two weeks. He offered to cut his gold Edison Medal into two pieces and give one to each. They declined, lending him money. It turned out that Tesla did have $5 in his office, but this money was reserved for bird seed.

When Tesla left the St. Regis, he moved with his birds to the Hotel Pennsylvania. After another few years he was forced to move to the Hotel Governor Clinton. Tesla and his pigeons spent the last days of their life in the Hotel New Yorker.

More On Women

Tesla had admired women from afar during his life, although he never had intraspecies sex as far as I can tell. When women in the 1920s started to become liberated and compete with men, he remarked "civilization itself is in jeopardy." This is not to say he didn't respect women. In fact, he believed they would eventually be intellectually superior to men. He said:

> The struggle of the human female towards sex equality will end in a new sex order, with the female as superior. The modern woman, who anticipates in merely superficial phenomena the advancement of her sex, is but a surface symptom of something deeper and more potent fermenting in the bosom of the race. It is not in the shallow physical imitation of men that woman will assert first their equality and later their superiority, but in the awakening of the intellect of women. Through the countless generations, from the very beginning, the social subservience of women resulted naturally in the partial atrophy or at least the hereditary suspension of mental qualities which we know the female sex to be endowed with no less than men. But the female mind has demonstrated a capacity for all the mental acquirements and achievements of men, and as generations ensue the capacity will be expanded; the average woman will be as well educated as the average man, and then better educated, for the

dormant faculties of her brain will be stimulated to an activity that will be all the more intense and powerful because of centuries of repose. Women will ignore precedent and startle civilization with their progress.

Tesla prophesied that society would soon resemble a beehive with "desexualized armies of workers whose sole aim and happiness in life is hard work. The acquisition of new fields of endeavor by women," he said, "their gradual usurpation of leadership will dull and finally dissipate feminine sensibilities, will choke the maternal instinct, so that marriage and motherhood may become abhorrent as human civilization draws closer and closer to the perfect civilization of the bee." There will be a "socialized cooperative life wherein all things including the young are the property and concern of all. The household's daily newspaper will be printed 'wirelessly' in the home during the night." I'm certain Tesla would have enjoyed today's Internet.

Tesla at Age 75

When Tesla was 75 and living in the Hotel Governor Clinton he tried to disprove Einstein's theory of general relativity. Tesla told interviewers that his own ideas were simpler than Einstein's, and when Tesla was ready he would reveal them to the world. He was a lifelong supporter of the physics of Sir Isaac Newton against the innovations of newcomer Einstein.

The aging Tesla was also working on a new source of power. "When I say a new source," Tesla said, "I mean that I have turned for power to a source to which no previous scientist has turned, to the best of my knowledge. The conception, the idea when it first came upon me was a tremendous shock." Tesla believed that the new power source would explain many unexplained cosmic phenomena. He also said it would be very valuable, "particularly in creating a new and virtually unlimited market for steel." Today, we still do not know to what Tesla referred.

Tesla also had a plan for transmitting energy between planets. "I think nothing can be more important than interplanetary communication. It will certainly come some day and the certitude that there are other human beings in the universe, working, suffering, struggling like ourselves, will produce a magic effect on mankind and will build the foundation of a universal brotherhood that will last as long as humanity itself."

Tesla's Strangest Invention

In 1933, when Tesla was 78 years old he described a device to reporters for photographing thought. "I expect to photograph thoughts," he said. "In 1893, while engaged in certain investigations, I became convinced that a definite image formed in thought must by reflex action produce a corresponding image on the retina, which might be read by a suitable apparatus. This brought me to my system of television which I announced at that time. . . . My idea was to employ an artificial retina receiving an object of the image seen, an optic nerve and another retina at the place of reproduction . . . both being fashioned somewhat like a checkerboard, with the optic nerve being a part of each." Tesla never revealed more details about this machine, which is illustrated in Figure 13.

Surtax of the Brain

In 1934, Tesla spoke of blood clots and brain atrophy that led him to have difficulty in concentrating. He continually tried to

> drive out of the mind the old images which are like corks on the water bobbing up after each submersion. But after days, weeks

13. Tesla's Thought Photograph Machine.

months of desperate celebration, I finally succeed in filling my head chockfull with the new subject, excluding everything else, and when I reach that state I am not far from the goal. My ideas are always rational because I am an exceptionally accurate instrument of reception, in other words, a seer. But be this true or not I am always mighty glad when I get through for there can be no doubt that such a surtax of the brain is fraught with great danger to life.

Tesla continued to believe in his idea of humans as "meat machines." He believed his meat machine concept was "one with the teachings of Buddha and the Sermon of the Mount." Our universe is

simply a great machine which never came into being and will never end. The human being is no exception to the natural order. Man, like the universe, is a machine. Nothing enters our minds or determines our actions which is not directly or indirectly a response to stimuli beating upon our sense organs from without. Owing to the similarity of our construction and the sameness of our environment, we respond in like manner to similar stimuli, and from the concordance of our reactions, understanding is born. In the course of ages, mechanisms of infinite complexity are developed, but what we call soul or spirit is nothing more than the sum of the functions of the body. When this function ceases, the soul or the spirit ceases likewise.

Tesla in the 21st Century

Tesla made many predictions regarding the future of humanity. He believed that by the year 2100 the world would uphold the tenets of a Buddhism/Christianity blend. By 2035, a Secretary of Hygiene would be more important than a Secretary of War. (He may be right, considering viral diseases like AIDS and the spread of antibiotic-resistant bacteria.)

In the 21st century, the world would spend more money on education than on war. Although Tesla originally thought wars could be stopped when countries grew afraid of terrible new weapons, he changed his mind. "I found that I was mistaken. I underestimated man's combative instinct, which it will take more than a century to breed out. War can be stopped, not by making the strong weak, but by making every nation, weak or strong, able to defend itself."

Tesla believed that ships would one day be propelled across the ocean by means of electrical currents projected from power plants on shore. As alluded to in a previous section, he also predicted that such currents passing through the upper atmosphere would allow humans to live in perpetual daylight. Plants providing the power would reside at Bermuda and the Azores. I wonder if this idea of perpetual light held any real appeal to Tesla considering he was a scotophiliac. Also, did he consider the affect of such proposals on the ecology of the world?

Tesla at Age 83: The Nabisco Cracker King

If we were to go back in time and visit Tesla at age 83, we'd find a man living almost entirely on milk and Nabisco crackers. In his hotel, Tesla numbered each cookie canister and stacked them neatly on his shelves.

Take a close look at Tesla, but don't go too close. He is so afraid of germs that all visitors must stand far away from him. Tesla is weak and asks a friend to care for his birds. Each day the friend spreads bags of seed near the New York Public Library. If he finds injured birds he brings them back to the Hotel New Yorker, where Tesla lives. Occasionally they give the pigeons a curtained shower bath.

Listen closely. During this last few years of his life, Tesla is suggesting that the United States build a "Chinese Wall" of his "teleforce rays" around the United States. He is telling you that the forcefield can melt airplanes 250 miles away. If the government would only give him two million dollars, he would develop the forcefield within three months. The War Department, however, declines his offer.

During the past few years, Tesla became especially interested in publicity and was a favorite character of newspaper interviews. For example, on July 11, 1934, the *New York Times* ran a story about Tesla's proposed death-rays. The headline was:

TESLA AT 78 BARES NEW "DEATH-BEAM"
Invention Powerful Enough to Destroy 10,000 Planes
250 Miles Away, He Asserts
DEFENSIVE WEAPON ONLY
Scientist, In Interview, Tells of Apparatus
That He Says Will Kill Without Trace

The proposed beam appears to have been a radio-wave scalar weapon or an ultrasound gun. In the 1940s, Tesla was even the subject of comic books. The first Superman cartoon, published in September 1941, features Tesla in *The Mad Scientist*. In this story a crazed, eccentric Tesla battles Superman using a death-ray that is also used to terrorize New York City.

Tesla Dies

A year after Enrico Fermi split the atom—a year after the first U.S. automatic computer was developed, a year after magnetic recording tape was invented—two influential, creative minds died. One was George Washington Carver, the famous agronomist. The other was Tesla. Neither Tesla nor Carver got to see the two popular films of that year: "Jane Eyre" (Orson Welles, director) and "Shadow of a Doubt" (Alfred Hitchcock, director).

On January 7, 1943, Tesla died at the age of 86, probably of a heart attack. After not hearing from him for several days, a maid decided to ignore the "Do Not Disturb" sign on his Hotel New Yorker door. Slowly creeping into his dark room, the maid discovered his emaciated corpse in bed.

Tesla's funeral service was on January 12 at the Cathedral of St. John the Divine, and his body was cremated. Tesla left behind huge numbers of research notes, but he left no will. Ten years prior to his death, Tesla stored a vast array of papers in the basement of the New Yorker Hotel. After his death the papers were examined by Dr. John G. Trump, a technical aide to the National Defense Research Committee of the Office of Scientific Research and Development. One of the documents he found was titled *Art of Telegeodynamics, or Art of Producing Terrestrial Motions at a Distance*. It described a device for transmitting mechanical vibrations from a machine bolted to a rock protruding from the Earth.

In a vault at the Hotel Governor Clinton was a device that Tesla had warned was a secret weapon that would detonate if opened by an unauthorized person. In the 1930s, Tesla gave the device to hotel management in lieu of rent money. He told them the device was very dangerous and worth $10,000. When federal agents arrived at the Hotel Governor Clinton after Tesla's death, they were handed this "dangerous" device wrapped in brown paper and tied with a string. They carefully removed the paper and found inside a polished wooden chest bound with brass. Inside the chest was a multidecade resistance box for measuring resistances—a common electrical device known today as a Wheatstone Bridge. This was not a new, unknown invention. To this day we do not know why Tesla wanted to terrify the hotel management with this harmless object.

Reopening the Tesla File

Although the FBI closed its Tesla files in 1943, they were forced to reopen them in 1957 after a couple living in New York started distributing newsletters containing information on flying saucers. The couple claimed that Tesla's engineers, after his death, created a "Tesla Set" for radio communications with UFOs. The FBI decided no action was warranted and closed the Tesla file for good.

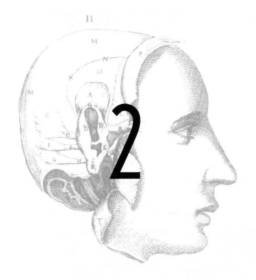

THE WORM MAN FROM DEVONSHIRE

IF IT IS LOVE THAT MAKES THE WORLD GO ROUND, IT
IS SELF-INDUCTION THAT MAKES ELECTROMAGNETIC
WAVES GO ROUND THE WORLD.

OLIVER HEAVISIDE, 1904

IT IS AS UNFAIR TO CALL A VECTOR A QUATERNION
AS TO CALL A MAN A QUADRUPED.

OLIVER HEAVISIDE, 1892

WHAT DREADFUL WORK CUTTING GRASS IS, WITH A
SCISSORS. 90 FEET BY 12, SAY. HARD ON
THE BACK, VERY.

OLIVER HEAVISIDE, 1897

🗁 FACT FILE

- **Name:** Oliver Heaviside
- **Born:** May 18, 1850, Camden Town, London, England
- **Died:** February 3, 1925, Devonshire
- **Occupation:** Inventor
- **Achievement:** A final contender for the 1912 Nobel prize, Oliver Heaviside established mathematical foundations for modern electric-circuit design and vector analysis for electromagnetics. His electrical theories allow us today to enjoy long-distance telephoning. Heaviside also predicted the existence of "superluminal radiation" of charged particles. Today this phenomenon is known as Cerenkov radiation. This occurs when a charged particle exceeds the speed of light in a dense medium, creating an electromagnetic shock wave shaped like a cone. Curiously, although physical theory forbids a particle from traveling at the speed of light in a vacuum, a particle violates no law if it travels faster than the speed of light.
- **Marital status:** Never married; celibate. His only long-term contacts with women were with his mother, nieces, and housekeepers.
- **Notable physical features:** Slim build, short (five feet, four inches tall), thick sandy brown hair, dark beard, piercing eyes, ruddy complexion. "Gentlemanly-looking."
- **Some bizarre behaviors:** Heaviside kept a female housemate as a virtual slave. Additionally, he exhibited scotophilia (loved the dark), claustrophilia (enjoyed working in tightly shuttered rooms), and thermophilia (loved to work in swelteringly hot rooms). Heaviside painted his nails a glistening cherry pink and replaced his comfortable furniture with granite blocks.
- **Frivolous signature:** Later in life he signed "W.O.R.M." after his name, signifying his low self-esteem. (Contrast this with Tesla's "G.I." signature, standing for "Great Inventor.")
- **Hobbies:** Music. Heaviside played the piano, aeolian, and pianola so loud, due to his hearing loss, that listeners would be in pain. He was also a great gymnast and bicyclist. "I really think the bike is a noble thing for sedentary people," he said.
- **A famous review of Heaviside's book, Electromagnetic Theory:**

  > "May the fact that I cannot understand 19/20 of your *Electromagnetic Theory* prevent me from congratulating you on the completion of Vol. II?" (Alexander Trotter, editor, *The Electrician*)

- **Religion:** A Unitarian, but not religious. Poked fun at those who put their faith in a Supreme Being.
- **Bird-love:** Heaviside, like Tesla, had difficulty forming relationships with humans, but loved birds. "A little bird has made friends with me," Heaviside wrote. "He knows what horrid creatures men are." Heaviside also kept a pet canary and once called himself the "man of many sparrows."
- **Favorite quotes about Heaviside:**

    > "Oliver Heaviside was an eccentric among eccentrics . . . by virtue of his talent and sheer force of will, he became one of the leading Victorian physicists. The next time you make a long-distance call and the voice on the other end comes through loud and clear, reflect for a moment on the gifted yet flawed man who made it possible. " (Paul Nahin, *Oliver Heaviside*)

    > "Heaviside's profound research into electromagnetic waves has penetrated further than anybody yet understands." (Oliver Lodge, *Nature*)

    > "The ability to follow Mr. Oliver Heaviside in his solitary voyages 'on strange seas of thought' is given to few. Most of us do get but glimpses of him when he comes into some port of common understanding for such fresh practical provisions as are necessary for the prosecution of further theoretical investigation. These obtained, he streams fast to sea again. Some of us in our puny way paddle furiously after him for a little distance, but we are rapidly left astern, and, exhausted, laboriously find our way back to land through the fog created of our own efforts." (*The Electrician,* 1903)

- **Powerful adversaries:** William Henry Preece (electrician and Engineer-in-Chief of the British General Post Office), Peter Guthrie Tait (Scottish physicist), and William Burnside (prominent mathematician)
- **Powerful supporters:** James Clerk Maxwell, father of electromagnetic theory, and George Francis FitzGerald, physicist
- **Little-known discoveries:** After radio waves spanned the Atlantic Ocean in 1901, Heaviside predicted the existence of a reflecting ionized region surrounding the Earth, which later became known as the Kennelly–Heaviside layer and is now called the ionosphere. Heaviside also invented a musical notation intended to be easier to read than the orthodox system of lines, bars, and notes.

- Prejudices: Heaviside may have been prejudiced against the working class. He writes, "I can't put up with common persons, wives of millhands and bricklayers and so forth; there are such a lot of vile blackguards going about here." He was also prejudiced against blacks. He once wrote, "Oliver Heaviside agrees never to marry a [expletive deleted]."
- Some quotes by Heaviside:

> "My principal infirmity is that I am a chronic dyspeptic, and it occasionally gets awful, and brings on nerve disturbance, which may culminate in epilepsy some day [like my mother], or may not, according as I live it down, or not. But this sort of thing is strictly for home consumption, most certainly not for public entertainment or otherwise." (1894)

> "The invention of quaternions must be regarded as a most remarkable feat of human ingenuity. Vector analysis, without quaternions, could have been found by any mathematician . . . but to find out quaternions required genius." (1892) [A quaternion is a four-dimensional number. Vector analysis is the study of quantities that have magnitude and direction.]

> "It goes so stiffly that I think I must have gout in my brain. . . . I have become as stupid as an owl."

 THE STRAIGHT DOPE

HEAVISIDE'S METHODS SEEMED A KIND OF
MATHEMATICAL BLASPHEMY, A WILLFUL SINNING
AGAINST THE LIGHT. YET HEAVISIDE'S RESULTS
WERE ALWAYS CORRECT! COULD A TREE BE REALLY
CORRUPT IF IT ALWAYS BROUGHT
FORTH GOOD FRUIT?

PROF. H. PIAGGIO
NATURE, 1943

Although he established the foundations of modern electric-circuit design, few people today have ever heard of Oliver Heaviside, the eminent Victorian mathematical physicist who made long-distance phone calls possible, who despised mathematical rigor, who was exces-

14. Oliver Heaviside circa (1891).

sively shy and eccentric, and who ridiculed his enemies in print. Given Heaviside's relative importance in the annals of science, it is amazing that only one or two complete biographies of Heaviside exist, my favorite being Paul Nahin's 1988 biography *Oliver Heaviside: Sage in Solitude*.

Heaviside was born in 1850—six years before Nikola Tesla—in a London slum. It was a year rich in ideas and inventions: Nathaniel Hawthorne's *The Scarlet Letter* was published; U.S. President Zachary Taylor died; and Millard Fillmore became the 13th president. During this time, the U.S. had 23 million people, 3.2 million of them slaves. The year's greatest scientific discovery was Herman von Helmholtz's determination of the speed of impulses in nerves.

Oliver Heaviside was the youngest of four sons. His father was a poor wood engraver who was always in bad health and probably physically abusive. The opening words to one of Heaviside's books on electromagnetic theory suggest that his father was a tyrant: "The following story is true,"

Heaviside wrote. "There was a little boy, and his father said, 'Do try to be like other people. Don't frown.' And he tried and tried but could not. So his father beat him with a strap. . . ." Later in life, Heaviside said, "When I was young, my father took me to the doctor in the hope of finding out whether it was what I had for breakfast that made me so stupid."

Although Heaviside's uncle was the famous Sir Charles Wheatstone, inventor of an electric telegraph, automatic transmitter, and the concertina, there is little evidence that Wheatstone contributed directly to Heaviside's education. However, at Wheatstone's suggestion, Heaviside added German and Danish to his language studies. Wheatstone was a man of many interests, including music, and it's not surprising that Heaviside soon developed a lifelong interest in music. Interestingly, today Wheatstone is remembered for the "Wheatstone Bridge," an electrical device for measuring electrical resistances—the same device that Tesla hid in a package in a hotel basement and was later discovered by police.

Heaviside's youth was an unhappy one, not only because of his father's abuse, but also because Heaviside's bout with scarlet fever left him permanently hearing impaired. This disability separated him from friends of his age, and that led to his aggressive and confrontational personality later in life. Heaviside himself said that his childhood problems left him "permanently deformed."

Like Nikola Tesla and Isaac Newton, Heaviside did not get along very well with boys his own age. Of his childhood, Heaviside remarked:

> I was born and lived 13 years in a very mean street in London, with the beer shop and baker and grocer and coffee shop right opposite, and the ragged school just around the corner. Though born and raised up in it, I never took to it, and was very miserable there, all the more so because I was so exceedingly deaf that I couldn't go and make friends with the boys and play about and enjoy myself. And I got to hate the way of tradespeople, having to fetch the things, and seeing all their tricks. The sign of the boozing in the pub made me a teetotaler for life. And it was equally bad indoors. A naturally passionate man, [my father], soured by disappointment, always whacked us, so it seemed. Mother similarly soured by the worry of keeping a school. Well, at 13, some help came, and we moved to a private house on a private street. It was like heaven in comparison and I began to live at once.

Heaviside was almost entirely self-taught. Although he went to school until age 16 and finished fifth among the more than 500 candidates for a College of Preceptors Examination given in 1865, he had no formal education after this point. Heaviside soon taught himself Morse code. With the help of his prominent Uncle Wheatstone, in 1870 Heaviside became a telegraph operator in Denmark. It was the only job Heaviside ever held, and in 1874, at the age of 24, he quit his job to devote himself entirely to private study. The first five years of his early retirement were his most productive: He wrote three important papers on the theory of the electric telegraph.

Oliver Heaviside's scientific papers were extremely difficult to understand because he never expended the effort to make them readable. "Fault has been found," Heaviside wrote, "with these articles that they are hard to read. They were harder, perhaps, to write." Certain readers considered his long equations with no explanation "a superior form of gibberish." Heaviside admitted that ordinary humans had to struggle to understand his work. "Let us now dig something out of the above formulae," Heaviside wrote. "This arithmetical digging is dreadful work, only suited for very robust intellects."

For the remainder of his life, Heaviside was supported by his brother, friends, and a government pension. Despite the fact that he had no advanced education, his genius in physics allowed him to make important contributions to theoretical and practical electromagnetics. His family, of course, was shocked and dismayed by his decision to forfeit paying jobs, but they did all they could to support him—even by leaving trays of food outside the door to his tightly shuttered room.

Heaviside's fondness for shuttered, dark rooms reminds me of Tesla's scotophilia. But Heaviside soon went beyond scotophilia and eventually became thermophilic—he loved working in swelteringly hot rooms by the light of smoky oil lamps. His friends called his room "hotter than hell."

Reminiscent of Tesla's decision to absorb the immense volumes of Voltaire no matter how daunting the task, Heaviside vowed to tackle James Clerk Maxwell's massive tome *Treatise on Electricity and Magnetism*. In 1918, Heaviside wrote about his obsessive need to absorb Maxwell's writings:

> I saw that it was great, greater and greatest, with prodigious possibilities in its power. I was determined to master the book and set to work. . . . It took me several years before I could

understand as much as I possibly could. Then I set Maxwell aside and followed my own course. And I progressed much more quickly.

Amazingly, Heaviside was able to condense Maxwell's equations with 20 variables down to just two equations in two variables. George Francis FitzGerald of Trinity College in Dublin wrote of Heaviside's achievements: "Maxwell's treatise is cumbered with the debris. . . . Oliver Heaviside has cleared these away, has opened up a direct route, has made a broad road, and has explored a considerable trace of country."

Since Heaviside's brilliant ideas on long-distance cable transmission were not in line with the views of established technical experts, W. H. Preece and H. R. Kempe, this resulted in Heaviside's exclusion from technical journals. Heaviside didn't take this quietly, and he had sarcastic, violent exchanges in print with these men. It turned out that Heaviside was right to suggest engineers add "induction loading coils" at regular intervals along telephone wires for long-distance, distortionless transmission of telephone messages, but he received neither payment nor much recognition for his ideas. However, Heaviside is indeed responsible for the coming explosion of telegraph and telephone use illustrated in Figures 15 and 16. Figure 17 shows his landmark article, published in 1887, announcing the distortionless phone line.

There was also an academic war over Heaviside's use of vectors describing forces as directional magnitudes. Even though Heaviside invented a whole branch of mathematics called operational calculus,[1] at the time his papers were rejected because of lack of rigor. Heaviside admitted he liked to skip steps and not get bogged down in all the details: "Well what of that?" Heaviside wrote. "Shall I refuse my dinner because I do not fully understand the process of digestion?" Later Heaviside wrote extensively on mathematical rigor:

> I appreciate the beauty of mathematical theorems occurring in Physics. I generally dislike very much the way they are "proved," as they say. Most mathematical books are a hodgepodge of formulas, without distinct connection to make a theory, or to exhibit it plainly, and made as repulsive and unintelligible as a legal document by attempts to be too precise and perfect.

Despite heavy opposition, the rest of the world gradually began to appreciate Heaviside's genius. His use of vectors in the world of electro-

15. Telegraph and telephone wires in Philadelphia (1890).

magnetics gradually replaced existing formulations, and textbooks eventually rewrote Maxwell's work in terms of vectors. Most scientists today believe that Heaviside was decades ahead of his time when it came to mathematical tools and notations.

Heaviside's brilliant mind wandered far and wide, and he did not confine himself to the study of electricity. After radio waves spanned the

16. Long-distance telephony (late 1800s).

Atlantic Ocean in 1901, he predicted the existence of a reflecting ionized region surrounding the Earth, which later became known as the Kennelly–Heaviside layer, now called the ionosphere. Heaviside's last known scientific work was in the late 1890s and concerned the age of the Earth as calculated from the time it takes for the Earth to radiate heat. After this research, Heaviside never published again in the technical journals.

Like Tesla, word of Heaviside's genius spread in his own lifetime, but in 1908 Heaviside became a hermit and moved to the southern coast of England. There in his small seaside cottage his neighbors treated him as a lunatic; most were unaware of his numerous honors.

Heaviside began to sign even his serious correspondence with "W.O.R.M.," signifying his low self-esteem and suggesting what his

Brothers, contractors for the electric lighting of a part of Turin, did not give the results that were expected in the public service.

"For the illumination of private parties, the contractors, instead of Gaulard transformers, installed the dynamos and transformers of Messrs. Zipernowsky and Déri.

"We have been informed that a large number of Gaulard transformers were found burned out and unfit for use, and Messrs. Bellani have decided to discontinue their employment for the public lighting. At the present moment the public service is secured by the Zipernowski-Déri system.

"The public illumination includes 80 incandescent lamps of 50 candle-power, and 3 arc lamps, all of which are run by the Zipernowsky-Déri system."

In addition, I have in my possession the results of tests made upon the Siemens cables before, during, and after the trials, which show that the failure must be ascribed to some other cause, considering, too, the fact that both systems used alternate currents, not differing greatly in electromotive force.

As the note in question contained no reference as to cause, but confined itself to facts, which Mr. Pickering has not shown to be untrue, I think he has no grounds for complaint.— Yours, &c.,   J. W. LIEB.

THE FIRE AT THE OPERA COMIQUE.

TO THE EDITOR OF THE ELECTRICIAN.

SIR : I was an eye witness of the catastrophe that occurred last night at the "Théâtre National de l'Opera Comique." It was about the end of the first act in "Mignon" when one of the gas jets in the cross light sputtered owing to a shock which that particular cross light received, due to some stage manipulation. The moment the light reached a sort of net decoration, in close proximity to the cross light, the net caught fire like ignited gun cotton. The decorations being so near the gas lights are always very warm and dry, so that they burn like tinder when lighted. The following is a little sketch in cross section, which shows how the fire originated :—

A is the iron frame of the cross light suspended by ropes G. B is the main gas tube, which has at regular intervals jets C, D is a sort of wire gauze, and F shows the decoration in front of the cross lights, E represents the sputter which ignited the decoration F, and caused the loss of so many lives and damage to property.

In a few moments the fire spread over the whole stage, and to-day the Opera Comique of Paris comes within the long list of such calamities which have occurred within the last few years, and reminds one of the Ring Theatre, Stadt Theatre in Vienna, the Arad Theatre and Szegedin Theatre in Hungary, the Nice Theatre, Rouen Theatre in France, and Alhambra in London. These are only a few of the principal ravages that our contemporary the gas light has been the cause of.

Had not the Cie. Continentale Edison been hindered from commencing their work last September, the disaster of yesterday would not have happened ; but instead of receiving their contract last year they only received it a few weeks ago. They have been actively pushing their installations, and expected to commence electric lighting in July.—Yours, &c.,
                                                        FRANCIS JEHL.
25, Rue Jacob, Paris, May 26th, 1887.

SECONDARY BATTERY QUESTIONS IN BELGIUM.

TO THE EDITOR OF THE ELECTRICIAN.

SIR : The statement which I have made as to the use of minium in the construction of accumulators rests upon the authority of the late Comte Du Moncel. When Du Moncel wrote the following paragraph there can be no doubt that his information came from a reliable source. He says :—

"Planté avait bien pensé, *des ses premières recherches*, à abréger le travail de la formation, en déposant sur les lames de plomb positives du minium et en réduisant cette couche par le courant polarisateur. . . . . mais l'adhérence de colle-ci était mauvaise et la couche déposée s'écaillait et finis sait par disparaître.

"Quand en 1881, on pensa à employer les accumulateurs pour l'eclairage électrique et les moteurs, M. Faure *reprit l'idée* de l'application de la couch de minium sur les lames de plomb, et pour maintenir l'adhérence, il enveloppa les lames recouvertes de minium dans des sacs de feutre." "Eclairage Electrique," Du Moncel, 1883, Vol. I., pages 56 and 57.

The problem which remained to be solved after M. Planté was merely to discover a satisfactory method of attaching the minium to the lead plate, and this problem, it is very certain, every one has a right to solve in his own fashion.—Yours, &c.,
                                                        YOUR CORRESPONDENT.

ELECTROMAGNETIC INDUCTION AND ITS PROPAGATION.—XL.

BY OLIVER HEAVISIDE.

(Continued from page 51.)

Preliminary to Investigations concerning Long-distance Telephony and Connected Matters.

Although there is more to be said on the subject of induction balances, I put the matter on the shelf now, on account of the pressure of a load of matter that has come back to me under rather curious circumstances. In the present article I shall take a brief survey of the question of long-distance telephony and its prospects, and of signalling in general. In a sense, it is an account of some of the investigations to follow.

Sir W. Thomson's theory of the submarine cable is a splendid thing. His paper on the subject marks a distinct step in the development of electrical theory. Mr. Preece is much to be congratulated upon having assisted at the experiments upon which (so he tells us) Sir W. Thomson based his theory ; he should therefore have an unusually complete knowledge of it. But the theory of the eminent scientist does not resemble that of the eminent scienticulist, save remotely.

But all telegraph circuits are not submarine cables, for one thing ; and, even if they were, they would behave very differently according to the way they were worked, and especially as regards the rapidity with which electrical waves were sent into them. It is, I believe, a generally admitted fact that the laws of Nature are immutable, and everywhere the same. A consequence of this fact, if it be granted, is that all circuits whatsoever always behave in exactly the same manner. This conclusion, which is perfectly correct when suitably interpreted, appears to contradict a former statement ; but further examination will show that they may be reconciled. The mistake made by Mr. Preece was in arguing from the particular to the general. If we wish to be accurate, we must go the other way to work, and branch out from the general to the particular. It is true, to answer a possible objection, that the want of omniscience prevents the literal carrying out of this process ; we shall never know the most general theory of anything in Nature ; but we may at least take the general theory so far as it is known, and work with that, finding out in special cases whether a more limited theory will not be sufficient, and keeping within bounds accordingly. In any case, the boundaries of the general theory are not unlimited themselves, as our knowledge of Nature only extends through a limited part of a much greater possible range.

Now a telegraph circuit, when reduced to its simplest elements, ignoring all interferences, and some corrections due to the diffusion of current in the wires in time, still has no less than four electrical constants, which may be most conveniently reckoned per unit length of circuit—viz., its resistance, inductance, permittance, or electrostatic capacity, and leakage conductance. These connect together the two electric variables, the potential difference and the current, in a certain way, so as to constitute a complete dynamical system, which is, be it remembered, not the real but a simpler one, copying the essential features of the real. The potential difference and the permittance settle the electric field, the current and the inductance settle the magnetic field, the current and resistance settle the dissipation of energy in, and the leakage conductance and

The article in which Heaviside announced the distortionless circuit.

17.   Heaviside's article announcing the distortionless circuit.

neighbors thought of him. He replaced his comfortable furniture with granite blocks. One of his neighbors writes:

> The granite furniture stood about in the bare room like the furnishings of some Neolithic giant. Through those fantastic rooms he wandered, growing dirtier and dirtier, and more and more unkempt—with one exception. His nails were always a glistening cherry pink.

Heaviside died in February 1925 from complications resulting from a fall from a ladder. It was the same year that John T. Scopes, schoolteacher, went on trial for violating the Tennessee law prohibiting teaching of evolution. Heaviside would never get to read the famous books published in the year of his death: F. Scott FitzGerald's *The Great Gatsby* and Franz Kafka's *The Trial*.

On the year of Heaviside's death, there was a solar eclipse in New York, the first in 300 years, a fitting tribute to a man whose keen eyes ranged from the stratosphere to the origins of the Earth, but who lived his last years in darkness.

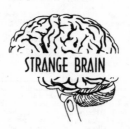

STRANGE BRAIN

HEAVISIDE: "I HOPE YOU DID NOT DRINK THE WATER IN THE BOTTLE UPSTAIRS."
PROF. SEARLE OF CAMBRIDGE UNIVERSITY: "NO, WHY?"
HEAVISIDE: IT HAS BEEN IN THAT ROOM SINCE THE LAST PERSON SLEPT THERE, THREE MONTHS AGO."

THAT WOMAN WANTED TO COME UP AND MASSAGE ME. I WANTED TO MURDER HER.
OLIVER HEAVISIDE, 1907

Author Richard Widdington in *The Post-Victorians* once wrote, "To the majority of people the unusual and arresting name of Oliver Heaviside conveys nothing." Why is it that so few people today have heard of this extraordinary genius whose ideas still affect our daily lives in the 1990s? How could a person with no formal education after the age of 16 come to be considered the intellectual superior of the finest minds of his day? These are questions not easily answered, but perhaps a cataloging of his unusual behaviors will shed some light. Certainly his eccentricity is one reason why many scientists of his day did not respect him and why his name was omitted from various scientific texts.

Oliver Heaviside lived a bizarre life and even held a woman his virtual captive for several years, as will be described later in this chapter. He loathed to be around strangers to such a degree that meeting a neighbor while walking near his home was considered a difficult, nasty task. He refused to work for money but accepted it from friends whom he often insulted even as they charitably gave him money. Aside from Heaviside's interest in painting his nails cherry pink and replacing his furniture with granite rocks, what more can we say of his odd compulsions?

Tesla and Heaviside had many similarities. They had little interest in women, were born to parents with few special abilities, and fell in love with electrical phenomena and birds. Like Tesla, who left behind boxes of obscure papers and notes after his death, Heaviside left sacks of papers covered with calculations that were discovered by authorities beneath the floorboards of his attic. Possibly Heaviside stuffed the papers there to insulate the attic.[2] In fact, the 1957 discovery of his "attic insulation mathematics" showed the world just how little Heaviside thought of his work. So detailed were the attic calculations that in 1959 H. Josephs published a paper suggesting Heaviside was working on a unified field theory. The paper starts with my favorite opening line from a scientific paper:

> On the 9th November 1957, Sir Edward Appelton received a letter which began "I have a sackful of Oliver Heaviside's papers in my garage . . ." [Josephs, H. J. (1959) The Heaviside papers found at Paignton in 1957. *Proceedings of the IEE.* 106C: 70–76.]

Despite Heaviside's many similarities with Tesla, Heaviside was mathematically much more adept than Tesla, whose work never involved sophisticated mathematics. Figure 18 shows a page from Heaviside's mathematical notes. Biographer Paul J. Nahin writes:

> Tesla was strictly an intuitive genius whose greatest insight, the rotating magnetic field, came to him in finished form (as far as he took it) literally "in his head." He neither used nor needed

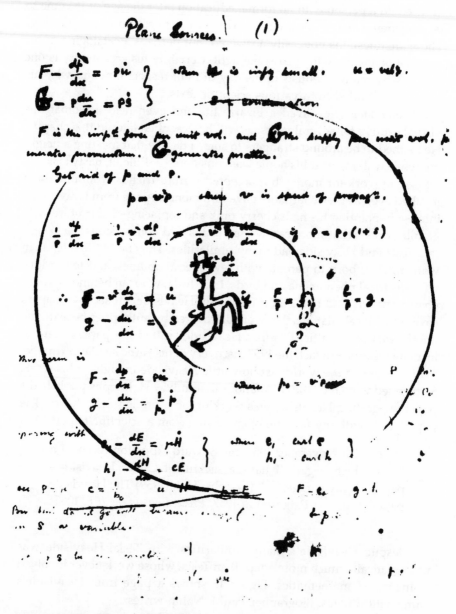

18. Wild side of Heaviside. Notice the vector pointing to the man's rear end in the mathematical doodle in Heaviside's mathematical notes.

analytic reasoning. If Tesla ever performed a mathematical analysis or displayed even the slightest comprehension of Maxwell's theory, I am unaware of it. . . . Tesla even claimed that crystals were "living beings."

In contrast, Heaviside's mathematical papers were so advanced that W. E. Sumpner said in his 1932 Kelvin Lecture:

In 1891, Heaviside summed up his work on Maxwell's theory in a single paper printed by the Royal Society in 1892. This was the most important and the most ambitious paper Heaviside ever wrote. It is fairly safe to say that no one yet born has been able to understand it completely.

Just how important was Heaviside's work? How can we compare his influence with other great inventors and scientists? One answer comes from Paul Nahin, who wrote, "Heaviside's work on how to make a decent telephone cable plays a vastly greater role in our everyday lives than does Einstein's work (because knowledge of the equation behind The Bomb, $E = mc^2$, is not essential for its making; i.e., Hiroshima and Nagasaki would have been doomed even had Einstein never lived)."

On Imagination

Heaviside wrote on many topics, from creativity and the soul to recipes for the perfect meal. Here is a brief excerpt from 1885 on the topic of imagination:

I am not objecting to use of the imagination. That would be absurd; for most scientific progress is accomplished by the free use of the imagination (though not after the manner of professional poets and artists when they touch upon scientific questions). But when one, by the use of the imagination, has got to a definite result, and then sees a stricter way of getting it, it is perhaps as well to shift the ladder, if not to kick it down. [Note: Ironic, considering Heaviside's death indirectly resulted from a fall from a ladder.] For I find that practically, in reading scientific papers, in which fanciful arguments are much used, it gives one great trouble to eliminate the fancy and get at the real argument. Nothing is more useful than to be able to distinctly separate what one knows from what only supposes.

On the Soul

Toward the end of his life, Heaviside wrote:

A part of us lives after us, diffused through all humanity—more or less—and through all nature. This is the immortality of the soul. There are large souls and small souls. The immortal soul of the "Scienticulists" is a small affair, scarcely visible. Indeed its existence has been doubted. That of a Shakespeare or Newton is stupendously big. Such men live the best part of their lives after they are dead. Maxwell is one of these men. His soul will live and grow for long to come, and hundreds of years hence will shine as one of the bright starts of the past, whose light takes ages to reach us.

``Brains All Smashed and Mixed Up''

Heaviside seems to have had a view of reality that I call "hyper-real"—he could write for pages on the minutest details of the world around him, almost as if he were observing the very fabric of his reality in a hallucinatory state. Excessive writing, or "hypergraphia," is a behavior often caused by certain forms of epilepsy that I will discuss in Chapter 16. Perhaps this type of minute analysis has value when applied to scientific investigations.

Heaviside's final years were spent in a seaside cottage where he became interested in flowers and birds. In 1897 he wrote a letter to his physicist friend George Francis FitzGerald:

There is a mysterious animal in the buttercup field, which has an oak in the middle. He begins at sunset and goes on for hours—perhaps all night; also Sunday morning, which is irre-ligious, but there must be a reason for it. I thought at first it was the beating of carpets whack-whack, whack-whack, very regu-lar, by two beaters. . . . It is really a mixture of a bleat and a croak. Trees about 150 yards away seems to be the place. They say it is a bird, probably a woodpecker, but I'll wait till I see it before I believe it. Then a horrid tramp came in and sneaked round my estate, seeking what he could devour. Moving from the room to watch his little game, a great horrid cat came and killed my canary. Then the tramp was spied by the gal, and forthwith begged for bread which he didn't get. No beggars allowed on my premises for bread or anything else, on principle. Bread indeed! I know better than that what they want . . .

[The greenhouse] is a rather cold place. . . . As I expected, I feel the change of climate. In fact, I took a dreadful chill, and it flew to the stomach and bowels as usual, and then to the brain, which was blown up over and over again. It is wonderful what the brain will stand; break and mend again. I have often wondered that I am not in a madhouse incurably imbecile, brains all smashed and mixed up. . . .

Later in life, like Tesla, Heaviside had difficulty forming relationships with humans, but loved birds. From the previous letter, we know Heaviside kept at least one pet canary. In addition, he enjoyed birds in his yard. At the risk of being repetitious, I include the following from Heaviside to give you an additional feel for his hypergraphia:

A little bird has made friends with me. Not a sparrow nor a robin, but something like a robin. Follows me about, and comes indoors and wanders about; especially present at meal time, when he comes in at the window and watches with interest what is going on, and picks up the crumbs that fall from the rich man's table. But he won't come near. . . . He knows what horrible creatures men are. But I wish he was more regular and less frequent in some of his operations. He makes his mark. It is dreadful to think that our ancestors perhaps went about doing the same. Evolution is most shocking in its early stages. Wonderful how things have worked out! If it wasn't true, no one could believe it. . . .

The following Heaviside letters should give you additional insight into his hyperdetailed world. Why should Heaviside expect Professor FitzGerald to care about Heaviside's surroundings in such minute detail?

June 19, 1897
Dear FitzGerald,

There is in the next garden an infant, terrible, and prodigious, with a most remarkably vociferous voice. "Dada, there's that, man, again! Dada! Dada! there's, that, man again! Dada, that, man, is going, into, the greenhouse! Dada! Dada! That, man, etc. etc." So I go and look over the wall to try if that row can't be stopped, and have a few words with the Dada about the weather and how nice his garden is looking, and so on. He is, I find, awfully deaf; conversation is only possible by shouting. That accounts for the extraordinary development of the infant's voice. He is a retired grocer, and is selling off his remainder of cheese on the premises at 7½ pence a pound. Had a pound, to soften *mores*. Perhaps you may have observed in your career that the judicious distribution of small sums of money has an emollient effect on some people. The British workman for instance. His moral character is instantaneously transformed by 6 pence especially if he hasn't earned it! However, as regards the grocer and his wife (who is also very deaf) the result was to make them inconveniently friendly. . . . She is one of those persons who go about the streets without a bonnet, of course I must draw the line somewhere; though I am decidedly democratic in principle, it is not always pleasant in practice.

The singular animal supposed to reside in the oak tree is still going on, with increased vigor. The blank space between his two croaks is shortened almost down to nothing. He is also heard, to the extent of an occasional few croaks, in the daytime. I think that is done in his sleep, as he begins regular work only at dusk. People going by at night might say "What a row!" and make imitative noises. Perhaps, however, it is a nocturnal phenomena of some kind, not an animal. . . .

There is a bat now flying about in my sitting room. It must have got in at the window upstairs, and then come down. Shall try and catch him; never examined one at close quarters. He goes round and round and may smash something before he has done. Not having any salt at hand, obliged to let him go. Opened front door, and he soon found the way out. But he went out too straight for a bat. Perhaps it was the old croaker out for a holiday. P.S. I hope you will not get drowned in crossing the Atlantic.

The following letter indicates Heaviside's unorthodox approach to maintaining his lawn:

July 12, 1897
Dear FitzGerald,

Wild horses were used to eat up the grass, so the "lawn" is, all over, holes, deep holes. It must be dug, or else filled up. What dreadful work cutting grass is, with a scissors. 90 feet by 12, say. Hard on the back, very. Lawn mower no good. Shears not good either, when the grass has to be lifted to cut it.

Heaviside on His Book

Many of Heaviside's contemporaries found his book *Electromagnetic Theory* difficult to understand. Alexander Trotter, editor of *The Electrician*, wrote, "May the fact that I cannot understand 19/20 of your *Electromagnetic Theory* prevent me from congratulating you on the completion of Vol. II?"

Heaviside admitted that his book was arcane in a letter to FitzGerald:

May 8, 1899
Dear FitzGerald,

As regards my Volume 2 of *Electromagnetic Theory*, well, I had a go at it Saturday and Sunday, having received a copy, and got a very bad headache. . . . It is a dreadfully dull book, unless you can go into it thoroughly; I was impelled to write it because someone or more than one must develop it. I really think it will in the long run have a considerable influence on the theory and practice of physical mathematics; still I cannot recommend any man of many cares and interests to trouble about it. I am sure it will not sell well. Volume 1 has sold nearly 600 copies. Volume 2 I expect say 300 in same time.

When asked by a book editor to provide a synopsis for a proposal for Volume 3 of his book, Heaviside wrote:

Synopsis? Can't. The Lord will provide. The best I can do is to suggest that you give me carte blanche and I will try to make the best use of it.

The editor replied that this would be acceptable!

The third volume of *Electromagnetic Theory* was published in 1912 and received a good review from *The Electrician*:

The new volume by Heaviside will prove a welcome addition to the libraries of mathematical electricians and physicists. Experience has proved that his writings will repay study, although many of us are often unable to soar with him in to the empyrean. It is a matter of history that he was the physicist who laid the foundations of the modern theory of telephonic transmission—a theory which has proved a veritable gold mine for the practical telephonist.

On Bikes

Heaviside frequently ran over chickens with his bicycle and considered bicycle exercise essential for a long life:

> I really think the bike is a noble thing for sedentary people. It can, of course, be made an instrument for athletic uses; you can toil at it if you like; but that is not necessary. Something was really wanted to take the place of the violent exercises indulged in a man's youth, they can't go on indefinitely; walking is too tame and tiring; you don't get any further. So most people settle down to hardly any exercise at all, and so degenerate . . . and go off too often in middle age—heart, liver, kidney and so on. The bike seems to me exactly the right thing to keep a man going on physically active to old age. I am delighted to see an old gentleman of 75 still at it. . . . Idiots consider me a madman about the bike; I ride every day. . . .

Figure 19 is a diagram of one of Heaviside's designs for a bicycle with decreased wind resistance.

On Food

Heaviside had very specific food preferences and an unnatural interest in food. He sometimes lived like a cat, drinking bowls or glasses of milk for days. Milk, and nothing else. (Strangely, Tesla also lived on milk, and for many years Thomas Edison's only foods were milk and the occasional glass of orange juice!) After poring through many of his known correspondences, I find that an inordinate number of them touch on food and food preparation.

19. Heaviside's bicycle design to decrease wind resistance. Heaviside, a great bicycle enthusiast, wrote of his invention: "It would not work as it could not steer. I am afraid one must raise the body above the level of wheels."

On potatoes and cauliflower, he once said:

I have had my domestic troubles too. Can't get a suitable house-keeper. So I do my own housework mostly. Simplified it considerably. Find that potatoes is the only food that can be eaten in quantity regularly. Easily cooked too. So when I get home at 1, I put on the potatoes immediately, then 1st course: one glass milk, one slice cake. Read paper. Then in 1/2 hour 2nd course: potatoes and butter. The butter is essential. . . . Sometimes I have a treat. A cauliflower.

On lentils:

The great lentil question cropped up today. Never! I had enough of it before. This learned girl had nuts for breakfast, because they were recommended by some idiotic vegetarian journal, and contained more nitrogen than anything else. Save me from nitrogen! It's a mad world. Lentils are High-Church. Always eaten in Lent by the stricter sort, as a penance, accompanied in private by hair-shirts and beads.

On pork:

Pork. Awful at night. Not indigestions, I think. Perhaps ptomaines [food poisoning]. Violent convulsions, body and brain, like a gymnotus [eel] in a passion.

On steak:

After your leaving, within a 1/2 minute, I was seized with a pain in the heart, which grew fast, and spread over a larger area. It was accompanied by a feeling of great anxiety, and of impending calamity, and despondency, and an aching desire for I know not what. I went and played Beethoven's Funeral March, but it did no good. Then came dinner: 2 oz of minced steak underdone was all I could eat. The potatoes might have been fine sawdust. Why, it was you and your wife sitting under a cocoa-nut tree in the island of Bermuda, she operating upon your ferocious moustache, and a lot of little naked [expletive deleted] boys looking on and grinning.

On grapes:

I have eaten a few grapes for 3 days past, right off the vine. They have an aperient effect. Is it due to the contents of the skins, or to the various poisons left on the outside of the skin by the various sorts of insects that crawl over it from time to time. In previous years I have noticed that only a minute or two after entering the greenhouse when the grapes are nearly ripe, I have been seized with the first symptoms of alcoholic poisoning!

On animal fat:

It is fat (real animal fat) that is needed in cold weather, rather than sugar. Ask the Greenlanders.

A more detailed account of his cooking and housework is given in his letters of 1899 when he talks about a housekeeper he brought from Paignton:

Middle aged Virgin had a stroke! Pretty piece of work. But she is getting over it nicely, and I think will be fit for work again. Doing easy work today, in fact. Caused by the cold; she was wretchedly clad (unclad, I should say).

After sending the woman away, Heaviside continued to cook for himself:

Quite independent, and have whatever I like for dinner. Stone broth, ditchwater soup. Made several discoveries. Parsnips cook easily. Carrots don't.

Heaviside smoked a pipe and abstained from alcohol. He once remarked, "I read with some surprise that the men in the Navy are served with a pint of grog (rum) a day. That is brutal. No wonder they behave so badly on shore. They should not have any at all."

In another letter:

I made some jam the other day out of some apples the boys had not stolen and some blackberries which I could not eat. But I am not fit for a cook. I forget. Then it all goes to cinder, to be discovered hours later. Or if I boil an egg, I am startled by a loud report, either I did not put any water in or else it has all boiled away.

In default of a proper working housekeeper I am inquiring for a respectable woman to come for the day, and then take her leave after tea. Even that is hard to find. I can't put up with common persons, wives of millhands and bricklayers and so forth; there are such a lot of vile blackguards going about here. I could tell such tales. But never mind . . .

On Health

Heaviside was forever writing letters on the status of his health. No detail was ever overlooked. Here are a few excerpts from letters to his friend Searle.

> I have been very ill since I saw you, and see no prospect of getting better till a great change in the weather takes place. Internal bleeding. Stomach and adjacent parts. The loss of blood is of no importance but it causes persistent diarrhea. Result of ulceration, I guess, and that's the result of repeated acute inflammations. It may go off. (March 5, 1909)
>
> Plague 5: Bunion bad, very. Large. Growing bigger. Soft. Painful. Walk on heel or side foot. Why not? Save swelling on ball of foot, of unknown cause, perhaps overeating if you don't like bunion. Wonder whether the rat poison would do it good. (January 11, 1913)
>
> Been very bad inside; ptomaine poisoning or something like. Dreadful pain all over intestinal tract, as they call it now. Then the liver very wrong, and violent Daggers. (January 24, 1913)
>
> Extraordinary variety of symptoms in past. Chalk in palms. Crystals from eyes. Tongue white, yellow, black, red. Stewed fruit nice, but makes wind. A new cauliflower for dinner is best, but can't be got.

On Neighbors and the Imprisonment of Miss Way

Heaviside frequently had trouble with his neighbors, who he said stared at him as if he were "an animal at the Zoo." He once wrote: "Boys in the field frequently calling [obscene names] . . . Sent note to Police. Didn't come. Spoke to Policeman. Didn't come."

Heaviside referred to his neighbors as "insolently rude imbeciles," and perhaps he was justified. They threw rocks at his windows and doors. His window was broken by a hunk of metal from a nearby street light. Boys would often plug the sewage drainpipe leading to his home. The surrounding area stank so much that it was mentioned in the local newspaper.

Things got so bad that his brother Charles made arrangements for Heaviside to stay with Mary Way, the unmarried sister of Charles' wife. Heaviside would live upstairs in her home for the last 17 years of his life.

Heaviside was brutal toward Way, keeping her as his virtual prisoner and slave, demanding that she not leave the house without his permission. Whenever possible, he prevented her from seeing friends and cut off her contact with the external world. He told her, "You must write to your friends and tell them not to come to see you." When she asked why, he responded, "Because you have got to do the work." Sometimes Way would leave the house without permission and not come back as soon as expected. When she returned, she often found Heaviside holding a lighted candle looking for her dead body in the garden.

During Christmas 1914, Heaviside's friends wanted to take Way to a concert, but she had no shoes in which she could walk comfortably. Heaviside would not let her go the two miles to Paignton for the help of a chiropodist.

Way once showed a friend a long contract Heaviside forced her to sign:

> Mary Way agrees never to marry a [expletive deleted]. Oliver Heaviside agrees never to marry a [expletive deleted]. Mary Way agrees to wear warm woolen underclothing and keep herself warm in winter. Mary Way agrees never to go out without Oliver Heaviside's permission. Mary Way agrees never to give anything away without Oliver Heaviside's permission.

Heaviside's thermophilia made him worry that Way was not wearing enough layers of clothing and therefore might catch a cold and die. He even asked a friend if she could check on Way's underclothing. Heaviside wrote in 1912, "My trouble has been that I can't warm the house, not till late in the evening, and the reason for that is the Baby's [Way's] persistence in leaving the doors wide open."

Way's imprisonment finally came to an end after seven or eight years. She had sank into a nearly catatonic state, spending all her time staring into Heaviside's fireplace fires. Her nieces finally came with a car, and without any warning, took her away leaving behind many belongings. Later Heaviside explained to others that Way had become "mad and had to be put away." (Recall that this was her house.)

Heaviside's thermophilia reached new highs as he required ever increasing quantities of gas to run both his lights and fires. He continually tried to improve the house's ancient piping. Once he opened and removed a main gas pipe, and lit the gas as it came roaring out. A huge flame shot out. Paul Nahin describes Heaviside's appearance. Friends who visited "were greeted by a blackened Heaviside peering at them, with one eye, through an opening in a blanket-bandage wrapped around his head and held in place by a rope tied around his neck!"

So strong were Heaviside's cravings for heat that he constantly fought with the local gas company about not paying his gas bills. In fact, he was using gas at the prodigious rate of 800,000 cubic feet per year. The following is a strange letter that he wrote to the gas company on the back of a large envelope from the Royal Society:

> From: Oliver Heaviside, W.O.R.M., "Wormfield", Torquay
>
> To: Manager, Torquay Gas Company
>
> Please send one new gas meter of strong constitution to replace the present one which is corroded both inside and outside by the rotten gas with which you are supplying me.

In 1924, Heaviside sometimes wore a tea-cozy on his head when his hair was wet, presumably to keep his head warmer.

The W.O.R.M. Man

> SEARLE IS ONLY A FELLOW OF THE SOCIETY OF WORMS, ENTITLED TO W.O.R.M. AFTER HIS NAME. AND I CAN TELL YOU THAT IT IS A GREATER HONOUR THAN F.R.S. WOULD YOU LIKE TO BE ENROLLED? THERE IS NO FEE.
>
> OLIVER HEAVISIDE, 1920

As previously mentioned, Heaviside signed his letters with the word "W.O.R.M.," indicating his obsession that the world despised him (Figure 20). "But I can't complain," he wrote, "being only a worm, to be crushed by barbarians and bullies." He also became obsessed with the very letters in his name, often writing them in alphabetical order. Numbers below each letter would indicate the number of times that letters appeared in "Oliver Heaviside." He wrote:

Home Field, Torquay. 8 Oct. 1922.

Dear Mr. & Mrs. Brown, and Miss Brown,

[handwritten letter in the hand of Oliver Heaviside]

Yours very truly,

Oliver Heaviside, W.O.R.M.

20. A letter from Oliver Heaviside with the title W.O.R.M. after his name.

My objection to medals and things of that sort has been a quite deep-seated one all my life, and the late enormous multiplication of so called "honors," by the million, removes the honor of distinction entirely. My W.O.R.M. title is the only one that I care for. It is quite a solitary distinction.

Heaviside's fear of people escalated with each year. In 1922, the Institution of Electrical Engineers decided to present Heaviside with its newest and highest award, the Faraday Medal. A deputation of four colleagues graciously agreed to come to Heaviside's home to present him with the honor. Heaviside wrote back:

> I don't know what you mean exactly by heading a deputation. Who are they? And I can't talk to more than one at a time, and that is not easy. It is brainwork mostly and eyework. . . . And I may not be able to get a room cleaned of the damn coal dust in time. Then you must *sit* in the hall or lobby. You must sit. It is very rude, to look down to a man. That's not "paying respect" to me is it? Nor is staring at me like the Zoo. Hadn't you better come 1 at a time on 4 successive days?

As if all of the above is not enough to indicate Heaviside's strange brain, consider that he kept a small bronze statue that he hung from the staircase by a swing. He would give it a push to make it swing like a pendulum as he stared at the back and forth motion of its shadow. He said, "That is the undying worm." He wrote to Searle, his Cambridge University friend,

> Dear Wormship and Lady Searle,
> You don't know it, but I was "the undying worm," long before you joined the Wormery.

The Death of the Worm Man

Near the end of his life, Heaviside really became quite ill. He had gout, swollen tonsils, and irregular heart beats. Heaviside never went to a doctor but treated himself by wrapping his body like a mummy in several layers of blankets from his upper chest to the top of his head. Even when he needed glasses, he would never go to a professional to obtain appropriate lenses.

In January 1925, friends could not get Heaviside to open the door to his home, and a few days later a policeman found Heaviside unconscious in his room and rushed him to a nursing home in an ambulance. Ironically, this was Heaviside's only ride in a motor vehicle.

The Worm Man from Devonshire

Heaviside died on February 3, 1925, and was buried by the grave of his parents. His death certificate listed several chronic ailments, but according to biographer Paul Nahin, falling 11 feet from a ladder and landing on his back is "more likely what did him in." Ten years later, the first round-the-world telephone conversation took place (two people in New York spoke by way of connections through San Francisco, Java, Amsterdam, and London).

Paul Nahin wrote:

> Heaviside should be remembered for his vectors, his field theory analyses, his brilliant discovery of the distortionless circuit, his pioneering applied mathematics, and for his wit and humor. The memory of what this enormously gifted man accomplished can only be blemished by the misguided appropriation of the work of others. Heaviside, a man of great personal integrity when it came to priority in technical matters, would have insisted that this is the only proper path to follow and anything else would simply be wrong. In our modern times, when the acquisition of fame, glory and material wealth have become consuming goals for so many, such integrity may mark Oliver as a distinctly strange and naive person.

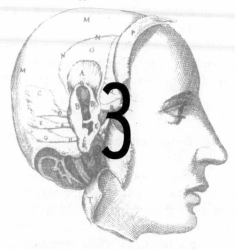

THE RABBIT-EATER FROM LICHFIELD

A MAN WHO USES A GREAT MANY WORDS TO
EXPRESS HIS MEANING IS LIKE A BAD MARKSMAN
WHO INSTEAD OF AIMING A SINGLE STONE AT AN
OBJECT TAKES UP A HANDFUL AND THROWS AT IT IN
HOPES HE MAY HIT.
SAMUEL JOHNSON

A MAN WILL TURN OVER HALF A LIBRARY TO MAKE
ONE BOOK.
SAMUEL JOHNSON

 FACT FILE

- **Name:** Samuel Johnson
- **Born:** Sept 18, 1709, Lichfield, England
- **Died:** December 13, 1784, Lichfield, England
- **Occupation:** Writer and lexicographer

- **Religion:** Zealous but joyless Christian churchman. "Johnson's fear of the Deity outweighed his love."
- **Strange brain:** Hypochondria, ate like a ravenous lion, auditory hallucinations, facial tics, elaborate rituals while entering doorways
- **Achievements:** His wide range of interests included science and manufacturing processes, and his knowledge seemed encyclopedic. Johnson is best known for his famous and witty books and essays, for example: *Dictionary of The English Language, The Lives of Poets, Miscellaneous Observations on the Tragedy of Macbeth, London, The Life of Savage, The Vanity of Human Wishes, Irene, Tatler, Spectator, Lay Monastery, Censor, Freethinker, Plain Dealer, Champion, Idler, Rasselas, Journey to the Hebrides, Taxation—No Tyranny.*
- **Favorite quotes about Johnson:**

  > "His mind resembled the vast amphitheater, the Coliseum at Rome. In the center stood his judgement, which, like a mighty gladiator, combated those apprehensions that, like the wild beasts of the Arena, were all around in cells, ready to be let out upon him. After a conflict, he drove them back into their dens; but not killing them, they were still assailing him. To my question, whether we might not fortify our minds for the approach of death, he answered, in a passion, 'No Sir, let it alone. It matters not how a man dies, but how he lives. The act of dying is not of importance, it lasts so short a time.' He added with an earnest look, 'A man knows it must be so, and submits. It will do him no good to whine.'" (James Boswell, 18th century)

- **Favorite quotes by Johnson:**

  > "Whatever withdraws us from the power of our senses; whatever makes the past, the distant, or the future predominate over the present, advances us in the dignity of thinking beings."

  > "Second Marriage: It is the triumph of hope over experience."

  > "The size of a man's understanding may always justly be measured by his mirth."

  > "The chains of habit are too weak to be felt until they are too strong to be broken."

🖙 THE STRAIGHT DOPE

S amuel Johnson (1709–1784)—the great scholar, poet, and witty playwright—is a prime example of another genius with obsessive–compulsive disorder. Although Johnson is not characterized as a scientist,
inventor, or philosopher, like the others in this book, his encyclopedic and
wide range of interests included science and manufacturing processes,
and his book *Journey to the Hebrides* is an important contribution to soci-
ology and anthropology.

Johnson was one of the most fascinating characters of the 18th century.
He wrote famous essays on Shakespeare and other poets and satire. His
monumental *Dictionary of the English Language* (1755) was the first general
dictionary of English. He fancied himself a brilliant man, and he surely
was one. In fact, an entire "age" was named after him. Johnson could
dominate almost any discussion, either with the depth of his literary
knowledge or by mere force of character. However, even after ignoring his
bluster and pedantry, Johnson displayed a great deal of wisdom regarding
life and human nature.

Young Johnson was sufficiently different from other children that his
parents took him to London when he was three years old to be inspected
by the court surgeon, prayed over by the court chaplains, and stroked with
a piece of gold from Queen Anne. Despite his weird facial contortions and
odd behaviors, he was a brilliant star in his classes. From age 16 to 18, he
lived at home, where he had no educational guidance. Nevertheless, he
ransacked his father's shelves and dipped into a never-ending stream of
books, reading whatever he found interesting. (His father was a book-
seller.) Once, while searching for a bite to eat, he found a huge volume of
the works of Petrarch, an Italian poet who lived from 1304 to 1374. The
name excited curiosity, and Johnson voraciously devoured hundreds of
pages in much the same way that Tesla absorbed the immense volumes of
Voltaire and Heaviside devoured of James Clerk Maxwell's massive tome
Treatise on Electricity and Magnetism.

Johnson was soon introduced to scholars who were amazed that,
despite his strange behaviors, he had absorbed an incredible quantity of
curious information gleaned while isolated in his father's library. At his
first day at Oxford, Johnson surprised his teachers by quoting Macrobius,
an obscure 5th century author. His teachers declared they had never
known a freshman of genius equal to Johnson.

When he was around 26 years of age, Johnson fell in love with a
widow twice his age. She was a short, overweight, crude woman who

dressed in gaudy colors. They married in 1735, and had a relatively happy marriage. In order to support themselves, Johnson decided to teach students in their home. Unfortunately, he only found three pupils because most were scared away by his contortions and twitchings.

When Johnson finally published his *Dictionary of the English Language*, it did not help him out of his poverty. The small amount of money the booksellers agreed to pay him had been advanced and spent before the book was published. Nevertheless, the world saluted Johnson as the highest authority on the English language. The *Dictionary* was "hailed with an enthusiasm such as no similar work had ever excited." The definitions were not simply dry tomes but rather showed a terrific command of the language with wonderful passages quoted from poets and philosophers.

In 1773, when Johnson was 63, he set forth on a journey to the Hebrides. Given his age, ailments, and purported opinion of the Scots, Johnson may have seemed a highly unlikely traveler to this distant region, but in the opening pages of his *A Journey to the Western Islands of Scotland* (1775) he confessed to a longstanding desire to make the trip. He was propelled by a curiosity to see strange places and study modes of life unfamiliar to him. His book, a superb contribution to 18th-century travel literature, combines important historical information with sociological and anthropological observations about the lives of common people.

When he was 72, Johnson's health deteriorated. He legs grew weaker. His asthma tortured him. For some reason, he urged surgeons to increasingly deepen incisions made in his body. So respected was Johnson that the ablest physicians and surgeons attended him and would accept no money from him. Although he was in a great deal of pain, every hour Johnson vehemently forced himself to pray. He finally died at 7:15 in the morning on December 13, 1784, and was laid to rest, a week later, in Westminster Abbey among eminent men.

STRANGE BRAIN

From childhood, Samuel Johnson suffered from several physical afflictions. By his own account, he was born "almost dead," and he early contracted scrofula (tuberculosis of the lymphatic glands). Because of a popular belief that the sovereign's touch was able to cure scrofula, he was taken to London at the age of 30 months and touched by the queen, whose gold "touch piece" he kept about him for the rest of his life. This ineffective remedy was followed by other medical treatments that left him with ugly scars on his face and neck. He was nearly blind in his left eye and exhibited very noticeable tics. Until he began to speak, new acquaintances sometimes took him for an idiot. Despite his many physical problems Johnson was vigorous, strong, and even somewhat athletic. He liked to swim, walk, and ride, even in later life. He was tall and became huge. A few accounts demonstrate his physical strength and strong character; for example, he hurled an insolent theater-goer together with his seat from the stage into the pit, and he once held off would-be robbers until the arrival of the police.

In order to achieve greatness, Samuel Johnson had to overcome the ravages of obsessive–compulsive disorder. For example, Johnson was an incurable hypochondriac. In addition, he could not step through a doorway without performing certain bizarre rituals. He never stepped on the cracks in paving stones, and he touched every post along the road as he walked. So compulsive was this behavior that if he missed a post, he would have to return to touch it while his friends waited. Johnson was always compelled to jump through doorways from a particular number of steps away. His friend, Miss Frances Reynolds wrote:

> Nor has anyone, I believe, described his extraordinary gestures and antics with his hands when passing over the threshold of a door, or rather before he would venture to pass through *any* door. On entering Sir Joshua's house with poor Mrs. Williams (a blind lady who lived with him), he would quit her hand, or else whirl her about on the steps as he whirled and twisted about to perform the gesticulations; and as soon as he had finished, he would give a sudden spring, and make such an extensive stride over the threshold, as if he were trying for a wager how far he could stride, Mrs. Williams standing groping about outside the door unless the servant or mistress of the house more commonly took hold of her hand to conduct her in, leaving Dr. Johnson to perform at the Parlor Door much the same exercises over again.

James Boswell, Johnson's noted biographer, also writes about Johnson's obsessive–compulsive disorder:

He had another peculiarity, of which none of his friends even ventured to ask an explanation. It appeared to me some super-stitious habit, which he had contracted early, and from which he had never called upon his reason to disentangle him. This was his anxious care to go out or in at a door or passage, by a certain number of steps from a certain point, or at least so as that either his right or his left foot (I am not certain which), should constantly make the first actual movement when he came close to the door or passage.

Johnson had a prodigious number of additional strange behaviors, some of which indicate Tourette's syndrome as much as obsessive–compulsive disorder. For example, his mutterings, grimaces, and gestures sometimes terrified people who did not know him. At the dinner table, he would absent-mindedly reach down and pluck a lady's shoe from her foot. For no reason at all, he would blurt out a fragment of the Lord's Prayer. Certain streets and alleys became places of horror for him and he would avoid them at all costs, even if it meant a long delay while traveling on foot. Sometimes he would look at clocks and not be able to tell the time. When he was a boy, he would distinctly hear his mother calling his name, even though she was miles away. Her voice was so real that he could not distinguish it from the real thing.

The sight of food sometimes drove Johnson into a wild frenzy and rapid orgy of eating—even in front of royalty. He had acquired the habit of eating like an animal. When he was near rabbit meat that had been kept too long, or a meat pie made with rancid butter, he would gorge himself with such violence that his veins swelled as the sweat poured from his head.

JOHNSON'S DICTIONARY

Despite Johnson's obsessions and compulsions, he achieved literary greatness. I recommend the 25-page Preface to the first edition of his *Dictionary* as an example of fine English prose. It also gives insight into Johnson's personality. The Preface begins:

Most men think indistinctly, and therefore cannot speak with exactness; and consequently some examples might be indif-ferently put to either signification: this uncertainty is not to be imputed to me, who do not form, but register the language;

who do not teach men how they should think, but relate how they have hitherto expressed their thoughts.

The *Dictionary* itself contains some amusing entries, for example:

fart. Wind from behind.
"Love is the fart
Of every heart;
It pains a man when 'tis kept close;
And others doth offend, when 'tis let loose."—Suckling
to fart. To break wind behind.
"As when we a gun discharge,
Although the bore be ne'er so large,
Before the flame from muzzle burst,
Just at the breech it flashes first;
So from my lord his passion broke,
He farted first, and then he spoke."—Swift

Here is another gem:

oats. A grain, which in England is generally given to horses, but in Scotland supports the people.

Johnson may have taken himself too seriously at times, but his humor is omnipresent. He was even humble at times. For example, consider the time he incorrectly defined "pastern" in his dictionary as "The knee of an horse." When he was asked by a lady why he made such a blunder, he replied, "Ignorance, madam, pure ignorance."

Johnson, like Oliver Heaviside, did not restrain himself from pillorying his enemies in print. For example, Lord Chesterfield turned Johnson away when Johnson needed money to work on his dictionary. After the dictionary was printed, Chesterfield praised the work and recommended people buy it. Johnson immediately wrote Chesterfield the following letter:

Seven years, my Lord, have now passed, since I waited in your outward rooms, or was repulsed from your door; during which time I have been pushing on my work through difficulties, of which it is useless to complain, and have brought it at last to the verge of publication, without one word of encouragement, or one smile of favor. Such treatment I did not expect, for I never had a Patron before . . .

Is not a Patron, my Lord, one who looks with unconcern on a man struggling for life in the water, and, when he has reached ground, encumbers him with help? [Johnson had defined "patron" in his Dictionary as "Commonly a wretch who supports with insolence, and is paid with flattery."] The notice which you have been pleased to take of my labors, had it been early, had been kind; but it has been delayed till I am indifferent, and cannot enjoy it; till I am solitary, and cannot impart it; till I am known, and do not want it. I hope it is not very cynical asperity not to confess obligations where no benefit has been received, or to be unwilling that the Public should consider me as owing that to a Patron, which Providence has enabled me to do for myself.

Having carried on my work thus far with so little obligation to any favorer of learning, I shall not be disappointed though I should conclude it, if less be possible, with less; for I have been long wakened from that dream of hope, in which I once boasted myself with so much exultation, My Lord, Your Lordship's most humble, Most obedient servant, Sam. Johnson

Dr. Johnson, like many great scholars, has many wise and witty quotes attributed to him. For example, when describing one of Shakespeare's minor plays, he said, "Criticism of unresisting imbecility is impossible." Here are some other favorites:

"There is a certain race of men that either imagine it their duty, or make it their amusement, to hinder the reception of every work of learning or genius, who stand as sentinels in the avenues of fame, and value themselves upon giving Ignorance and Envy the first notice of a prey."

"It is better to *live* rich than to *die* rich."

"We are perpetually moralists, but we are geometricians only by chance."

"My wife had a particular reverence for cleanliness and deserved the praise of neatness in her dress and furniture as many ladies do till they become troublesome to their best friends, slaves to their own bosoms and only sigh for the hour of sweeping their husbands out of the house as dirt and useless lumber."

"In the tumult of conversation malice is apt to grow sprightly."

"Let us be serious, here is a fool coming."

"Music is the only sensual pleasure without vice."

"No man but a blockhead ever wrote, except for money."

"No man will be a sailor who has contrivance enough to get himself into a jail; for being in a ship is being in jail, with the chance of being drowned."

On playing the violin: "Difficult do you call it, Sir? I wish it were impossible."

"Sir, this man has a pulse in his tongue."

"There is no being so poor and so contemptible who does not think there is somebody still poorer, and still more contemptible."

"Illiterate writers will at one time or another, by public infatuation, rise into renown, who, not knowing the original import of words, will use them with colloquial licentiousness, confound distinction, and forget propriety . . . But if the changes that we fear be thus irresistible . . . it remains that we retard what we cannot repel, that we palliate what we cannot cure."

"Your manuscript is both good and original; but the part that is good is not original, and the part that is original is not good."

"Read over your compositions and, when you meet a passage which you think is particularly fine, strike it out."

"Nothing will ever be attempted, if all possible objections must be first overcome."

"I never think I hit hard unless it rebounds."

"We are all agreed as to our own liberty; we would have as much of it as we can get; but we are not agreed as to the liberty of others."

"No man ever yet became great through imitation."

On the British government's actions in an 18th-century Falklands war: "Patriotism is the last refuge of a scoundrel."

When we read through the witty words of Johnson that seem to pour from him like water gushing from a fire hose, one wonders if these kinds of brief, concise aphorisms were a by-product of his disorders that pushed him to place everything quickly in their places. Samuel Johnson, English essayist, explorer, biographer, critic, poet, and lexicographer, once characterized literary biographies as "mournful narratives," and he believed that he lived "a life radically wretched." Yet his career is a success story of the sickly boy who by brainpower, talent, and tenacity became the foremost literary figure and the most formidable conversationalist of his time. For future generations, Johnson was synonymous with the later 18th century in England. The disparity between his circumstances and accomplishments gives his life its special interest.

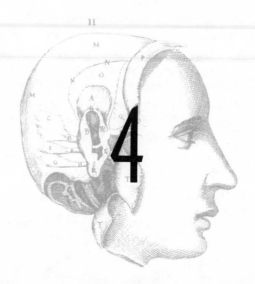

THE FLY MAN FROM GALWAY

 FACT FILE

- **Name:** Richard Kirwan
- **Born:** 1733, Cloughballymore County, Galway, Ireland
- **Died:** June 1, 1812, Dublin, Ireland
- **Occupations:** Chemist, mineralogist, geologist, meteorologist
- **Achievements:** Kirwan was internationally famous for his studies in mineralogy, logic, music, philology, meteorology, and law. He won Britain's Copley medal for chemistry and was made a Fellow of the Royal Society. His books include *Elements of Mineralogy, An Estimate of the Temperature of Different Latitudes, Essay of the Analysis of Mineral Waters,* and *Geological Essays.*
- **Strange brain:** Hypochondria, dysphagia, dipteraphobia (fear/hatred of flies), thermophilia, nocturnal camelliaphilia
- **Religion:** Catholic turned Protestant turned Unitarian
- **Favorite paper title:** "What are the Manures Most Advantageously Applicable to the Various Sorts of Soils, and What Are the Causes of Their

Beneficial Effect in each Particular Instance" (*Transactions of the Royal Irish Academy*, 1794).

☞ THE STRAIGHT DOPE

Richard Kirwan was born in 1733 to a wealthy family, and he displayed a remarkable intellect at a young age. Little is known about his early years except for the fact that his father died when Kirwan was eight years old. Other family members, including his mother, would soon follow Kirwan's father to an opulent grave.

Since Kirwan's Catholic religion prevented him from attending most British universities, in 1750 he enrolled at the University of Poitiers, where he displayed an unusual interest in chemistry at the exclusion of his other subjects. In France he was obliged to learn the French language, a task he disliked. However, his tutors noticed his interest in chemistry, and restricted his chemistry studies to French textbooks, which soon encouraged his fluency.

In 1754 Kirwan left the university to become a Jesuit novice. However, he soon found that the Jesuit discipline did not suit him, nor did his religion. (As a result he became a Protestant in 1764 and died a Unitarian.) When his older brother died in a duel, Kirwan inherited his family's luxurious estate, and he abandoned the Jesuits. He returned to Galway a "tall, elegant, comely young man, given to interceding his discourse with foreign idioms."

In 1757, he married Anne Blake, daughter of Sir Thomas Blake of Menlough Castle in Galway. Kirwan lived his married years in the home of the Blakes, where he built a laboratory for his personal research. Kirwan soon became interested in law, and in 1764 he renounced his Catholicism, in part, so that he could be called to the Irish bar. He practiced law for two years but found it rather dull and unrewarding. He then spent the next eight years increasing his knowledge of science and languages. When Kirwan was 32 years old, his wife died leaving him with two daughters. In 1788 he returned to Ireland, having completed 20 years of study in such fields as metaphysics, law, logic, philology, chemistry, mineralogy, philosophy, mining, geology, and meteorology. When he settled down in Dublin, he was not only recognized as an outstanding genius, but acquired a national reputation as a weather prophet.

Kirwan presented a large number of scientific papers to the Royal Irish Academy of which he was president from 1799 until his death. He was elected to the Royal Society in 1780 and awarded the Copley medal for his work in what he called chemical affinity. "Chemical affinity," he said, "or attraction is that power by which the invisible particles of different bodies intermix and unite with each other so intimately as to be inseparable by mere mechanical means." His experiments on the specific gravities and attractive powers of saline substances formed a significant contribution to the methods of analytical chemistry. Specific gravity, also called relative density, is the ratio of the density of a substance to that of a standard substance. The usual standard of comparison for solids and liquids is water at 4 degrees C (39.2 degrees F), which has a density of 1.000 kg per liter (62.4 pounds per cubic foot). Specific gravity is the basis of methods used to concentrate ores. Panning, jigging, shaking, spiral separation, and heavy-medium separation are among the ore-dressing methods that depend on differences in specific gravity to obtain concentrated ore.

Richard Kirwan's pioneering work in meteorology included a system of forecasting the weather for seasons ahead using records of Irish weather spanning 41 years. His predictions were useful to farmers, and his mathematical methods (autocorrelation) are used today in a variety of disciplines. For example, autocorrelation can be used to understand climatic variation. A long time series of a quantity such as temperature often displays autocorrelation; i.e., its present behavior is related to its behavior at other times. Some of these correlations are of fundamental importance. For example, the frictional drag of the winds on the sea surface, which partly governs the oceanic circulation, depends on the covariance (a measure of correlation) of vertical and horizontal components of the wind and on similar covariances within the sea.

Kirwan was a practical scientist, always wondering how his theories might be helpful to industry. He published on topics as diverse as coal mining, manures, and bleaching. In fact, Kirwan introduced chlorine bleaching to Ireland. Bleaches were used to whiten or remove the natural color of fibers, yarns, paper, and textile fabrics. His book *Elements of Mineralogy* was the first systematic work on mineralogy in the English language and has long remained a standard.

Despite Kirwan's pioneering work in mineralogy, he did have a few theories about the formation of the Earth that would sound odd to many scientists today. Like other theorists of his day, Kirwan believe that all rocks of the Earth's crust had been precipitated from some primordial flood. According to Kirwan, the flood originated in the southern hemi-sphere and swept northwards with "resistless impetuosity." Its surging waters reshaped the continents (e.g., Africa and South America) to give them their southward taper, shivered a primitive land-mass in the north Pacific to leave only a few islands, and then swept across Asia and North America, in some regions dashing mountains to pieces, and in others scouring the terrain to leave barren, soil-less deserts such as the Gobi. Kirwan believed that his theories were in accord with *Genesis*, but some of his colleagues thought otherwise and accused Kirwan not only of being deficient in field experience but of superficiality in his "reading of Moses." It may be that Kirwan took this latter point to heart, because we are told that during the closing years of his life he spent long periods immersed in Scriptures.

When he reached his 70s Kirwan lost interest in science and published on logic and metaphysics. In one of his papers, he attempted to prove that humankind's first language was ancient Greek. Kirwan finally died at 79, in 1812.

STRANGE BRAIN

Richard Kirwan's early genius was evident to his mother, who be-came alarmed at his extreme interest in books and study at the exclusion of the rural pastimes of other boys. When he was five, he conjugated French verbs; at seven he wrote an abridgement of ancient history. Growing increasingly alarmed by the intensity of his studies, his mother wrote him the following letter (Kirwan was 17):

> You see whether I have cause to be uneasy about it when I tell you the misfortune of two that were eminent in that way; one Furlong, who found out the way to make Bath metal grey, by study at last melancholy, let his beard grow and talked to himself. In short, by all I heard, he was lost by it; and the Dominican Friar . . . that was the end of their labors and profound studies, as they fancied. There are several instances of

people that were turned or touched, as they call it, by studies which make me insist so long upon your not falling into the dangerous practice. . . .

Richard Kirwan, the brilliant 18th-century polymath, was renowned for his conversation and his books on mineralogy and geology. What you won't find in your encyclopedia is the slightest mention of Kirwan's bizarre obsessions, compulsions, and throat. For example, Kirwan always ate his food in solitude. He suffered from dysphagia—chronic difficulty in swallowing—and he didn't want other people to see his contortions necessary to get food down his throat. For reasons I have not been able to determine, his entire diet consisted of only milk and ham. (Recall all the other strange brains who had a strong preference for milk.) The ham was cooked on Sunday and reheated every day for the rest of the week.

Kirwan's home contained little furniture and all the chairs and tables overflowed with books, which also lay in piles on the floor. Nevertheless, his home became a popular meeting place for famous scientists, debutantes, and rich people. After he retreated into a back room and underwent his convulsive movements to swallow his ham and milk, he would walk right back out to his company and became quite social.

Every Thursday and Friday, he invited guests for conversation and music starting at six o'clock p.m. and ending promptly at nine o'clock p.m. Guests better not arrive late because he removed the knocker to his front door at seven o'clock. If his guests did not leave at nine o'clock, Kirwan put on his pajamas and escorted them to the door. At this point in his life, Kirwan's guests described him as "a tall gaunt figure, wrapped from neck to heel in a dark roquelaure, with a large leaved hat flapped low over his face."

Kirwan had an obsessive hatred of flies and would pay people money for each one they could kill and bring to him. Once a worried group of civic dignitaries, including the Lord Mayor, came to Kirwan's home for advice on putting out a dangerous fire blazing in the city coal yard. Kirwan's trusted servant, Pope, came to the door and refused to let them in.

"If you were the king himself I dare not admit you," Pope said.

"The city coal yard is in flames!" they exclaimed.

"If all Ireland were in flames and this house along with it, I dare not; for my master is at dinner."

After much begging, the delegation was allowed in, but in doing so a few flies came through the open door.

Pope's screaming voice stopped the Lord Mayor. "Oh sir, they will get in!" Pope yelled.

"What!"

"The flies! The flies!" he shouted, waving a cloth.

Kirwan soon offered money to his servants for each fly corpse they could bring him. Finally, Kirwan gave the delegation sensible advice about putting out the fire with sand.

Although Kirwan was known for his generosity and charitableness throughout his life, he was quite testy when it came to reading letters on subjects that did not interest him. The letter I recall most vividly was one that he wrote in reply to his brother, "Dear brother," Kirwan wrote in a return letter, "I have read over twice the letter you have pleased to send me, which to me, who hates reading or writing or any business, was a very disgusting task."

Kirwan also was an extreme hypochondriac, like Heaviside. His primary concern was that he would die from catching a cold, and he took strange measures in an attempt to avoid colds. In particular, he thought he could store heat in his body like a gigantic thermos bottle. In order to avoid colds, his daily routine involved getting up at four o'clock and sitting in his drawing-room, which had a huge fire blazing in it all year round. Even in the summer, guests to his house would find him sitting in front of a roaring fire. His thermophilia, like Oliver Heaviside's, reached new heights when he decided to always wear a hat indoors in order to retain body heat.

21. Richard Kirwan's worst nightmare.

When he wanted to go out for a walk with his huge dogs (mastiffs, greyhounds, Irish wolfhounds) and tamed eagle, he would first have to absorb and store heat by sitting in front of a fire. Then he would walk briskly, keeping his mouth tightly shut to retain the heat. His eagle sat on his shoulder. He always preferred large creatures to small ones ever since the time his dogs saved his life when he was attacked by six boars while he slept. Kirwan had trained his "attack-eagle" to be loyal by starving it and employing a boy to make its life a misery by teasing it. Kirwan would then suddenly appear with a plate of meat and drive the boy off.

When Kirwan walked with friends, even the Viceroy, he always hurried along, obsessed with retaining the supply of heat, and they would have to run beside him if they wanted to keep up. He refused all conversation while walking.

Pope, Kirwan's servant, slept in Kirwan's bedroom and was instructed to wake him every few hours to pour hot tea down Kirwan's throat. Most likely by accident, Pope would often miss Kirwan's mouth and dump the tea on Kirwan's nose and hair making a horrible mess in bed. Sometimes he mistakenly poured the spout in Kirwan's eye, rather than his mouth. The purpose of all the nocturnal tea drinking (camelliaphagia) was to sustain Kirwan's internal body heat through the night. One wonders what Kirwan would have done with a modern electric blanket.

Kirwan finally died at the age of 79 from the very thing he feared: complications from a cold. (Kirwan tried to cure the cold by the fashionable method of starving, which probably didn't help matters.) When Pope died, he shared his master's grave.

Why did so many of the strange brains suffer from hypochondriasis, a mental disorder characterized by an excessive preoccupation with one's own health? Most hypochondriacs fear or believe they have a serious disease based on insignificant symptoms. Some exaggerate the medical significance of minor aches and pains, becoming morbidly and obsessively preoccupied with the thought of a life-threatening illness. Their fears usually persist even after a physician has determined that no physical abnormality exists. As a result, a doctor's reassurances have little effect on such a person's apprehensions. As with many people with obsessive–compulsive disorder, the typical hypochondriac does not usually become delusional about his health. In other words, hypochondriacs often admit the possibility that their fears are unfounded. Despite this, many hypochondriacs will go from doctor to doctor in their effort to find a cure for an imagined illness.

Where do hypochondriacs usually report their greatest concerns? Most hypochondriacs focus on a supposed disease of the heart, gastrointestinal tract, or genital structures. They may become fixated on disorders as a result of reading the medical literature or current public health concerns and fads. Hypochondriasis may exist by itself or it may be a secondary syndrome that occurs along with another mental disorder. Hypochondriasis usually starts in the teen years and is equally common among males and females. Some believe that hypochondriasis is a psychological coping mechanism for a person dealing with stressful life situations. Sometimes psychotherapy can be helpful in determining an underlying emotional cause, and behavior therapy may also be helpful.

For years, hypochondriasis without accompanying depression was considered to be unresponsive to pharmacotherapy. However, recently Dr. B. A. Fallon and colleagues report in the *Journal of Clinical Psychopharmacology* that fluoxetine (Prozac) can be helpful, just as it is for obsessive–compulsive disorder. Hypochondriacs, in their obsessions about illness, compulsions to check with others, and failure to be reassured, share many features in common with those who have obsessive–compulsive disorder. Given the marked similarities between these disorders, researchers decided to conduct a trial of high-dose fluoxetine for patients with hypochondriasis who did not meet criteria for major depression. Ten of 16 patients were much improved at the end of 12 weeks. These results suggest that fluoxetine is a useful therapy for hypochondriacal patients without marked depressive features.

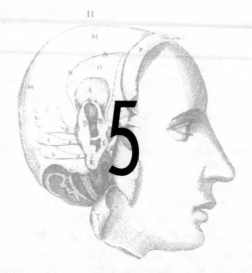

THE RAT MAN FROM LONDON

 **FACT FILE**

- **Name:** Jeremy Bentham
- **Born:** February 15, 1748, London
- **Died:** June 6, 1832, London
- **Occupation:** Philosopher, economist, and theoretical jurist
- **Accomplishments:** Through his disciples, James Mill and John Stuart Mill, Bentham became the father of the utilitarian school of philosophy. Bentham published various insightful books on government and economics, including *A Fragment on Government*, *The Rationale of Reward*, *The Rationale of Punishment*, *Defense of Usury*, *An Introduction to the Principles of Morals and Legislation*, *A Catechism of Parliamentary Reform*, and *Constitutional Code*. His *Catechism* advocated annual elections, a wide suffrage, equal electoral districts, and secret ballots.
- **Unusual accomplishment:** Bentham invented "felicific calculus," a method for calculating amounts of happiness. The method turned out to be im-

practical because users would have to be omniscient gods to include all the input variables.

- **Strange brain:** Extreme shyness and lifelong seclusion, rodentphilia, strange ideas about corpses, friendships with household items
- **Famous quotation:** His utilitarian ideal, "the greatest good for the greatest number of people," influenced the U.S. Constitution, angered despots, and inspired liberal revolutionaries during the late 1800s and early 1900s.
- **Marital status:** Celibate; never married

☞ THE STRAIGHT DOPE

Like many of the other influential geniuses discussed in this book, Jeremy Bentham showed early signs of brilliance. For example, he read a history of England at age three, and at five he read Greek and Latin. After he graduated from Oxford University at the age of 15, he reluctantly followed his father's suggestion to enter law school. While at law school, he listened to Sir William Blackstone, a future judge, and detected flaws in the judge's logic. As a result, Bentham spent his time speculating on theoretical aspects of legal abuses rather than practical law. To the disappointment of his lawyer father, Bentham never practiced law. Bentham's first book, *A Fragment on Government*, was published anonymously in 1776. It examined governments, criticized the English Constitution, and exhibited Bentham's indifference to historical justifications of current law. Lord Shelburne, a famous statesman, was so impressed with the book that he frequently asked Bentham to be a guest to his home.

Over the years, Bentham published a number of books on government and economics. In his *Manual of Political Economy*, he gives a list of what the state should and should not do, the second list being much longer than the first. In his book *An Introduction to the Principles of Morals and Legislation*, Bentham defines the principle of utility as "that property in any object whereby it tends to produce pleasure, good, or happiness, or to prevent the happening of mischief, pain, evil, or unhappiness to the party whose interest is considered." Humans are governed by pain and pleasure. The object of all laws should be to produce the "greatest happiness of the greatest number." All punishment involves pain and is therefore evil. Because of this, punishment should be used only "so far as it promises to exclude some greater evil." Through his disciples, James Mill and John

Stuart Mill, Bentham became the father of the utilitarian school of philosophy, which advocates the intellectual and social independence of individuals and defends civil liberties and democratic ideals.

Bentham was actually more concerned with improving society than with private morals. He wasn't a revolutionary but wanted people to work through parliamentary procedure. He advocated governmental supervision but not governmental control. Whenever asked his opinion on various governmental policies, his prime question was: "What is its use?" This test of usefulness was his major principle when deciding how to revise and codify law. (Did he take "utility" too far when he said that poetry was simply prose that failed to cover the page?)

The fame of his *Principles* spread faster than a best-selling novel. Because his reforms facilitated peace and public order, Bentham was made a French citizen in 1792, and his was advice sought throughout Europe and America. Bentham continued to be fascinated by legal systems, and his dream was to be allowed to prepare a code of laws for an entire country.

According to Bentham, only pleasure (and avoidance of pain) was the proper measure or criterion of right conduct because four types of penalties would punish excessive selfishness:

- Physical—overindulgence causes nausea.
- Political—imprisonment awaits those who infringe on others' rights.
- Moral or popular—public opinion ostracizes evil.
- Religious—God punishes antisocial pleasure.

Bentham's Hedonistic Calculus, or "felicific calculus," provided a means for computing the amount of pleasure an individual may anticipate. The calculus considered seven quantities of pleasure when quantifying happiness:

- Intensity of pleasure
- Duration
- Certainty—how probable is the expected pleasure?
- Propinquity or remoteness—how long must one wait for the pleasure?
- Fecundity—how many future pleasures result, and how many will end pain?
- Purity—is any pain mixed with the pleasure?
- Extent—how many others can share in your pleasure?

Although it may seem that this system places too much emphasis on individual happiness, Bentham believed that our separate joys are shared because the pleasures and interests of each one of us intersect with those of others.

Bentham was also active in prison reform and fervently tried to induce governments to accept his ideas. When he did not succeed, he lost faith in politicians.

STRANGE BRAIN

Jeremy Bentham is considered by some to be the most important mind in British history, yet he preferred the company of animals to humans. His mind may have been powerful, but his body was not. As a child, his feeble muscles forced him to creep up stairs like a slug. At the age of seven, he could not dance because his muscles were too weak. He couldn't even stand on his tiptoes. (I have not been able to determine the exact disease or ailment from which he suffered.)

Like Oliver Heaviside's father, Jeremy Bentham's father was domineering almost to the point of abuse. As just one example, consider that Bentham always addressed his father as "Honored Sir." To make matters worse, Bentham's father would gather friends around his little genius son and exhibit poor Jeremy for observation, like a circus animal. The shows would go on until little Jeremy became so embarrassed that he was "ready to faint—to sink into the earth with agony." Later, Bentham's father bullied Jeremy into entering law. His forced entry to law school was so horrifying that Bentham wrote, "I went to the bar as a bear to the stake."

To the bitter disappointment of his father, Bentham rejected the entire British legal system, calling it arbitrary and unjust, and began to formulate a more scientific legal system. For the next 50 years, Bentham went into seclusion, obsessed with his grand schemes for governmental reform and his principle of utility—that the purpose of all things should be to enhance human happiness and reduce human suffering. Every day, he wrote a dozen pages, most of which were never finished and never published. Rats began to nibble on most of his work. Bentham's friends and disciples practically rewrote several of his books from mounds of tattered memoranda.

Bentham manifested his antisocial behavior in many ways. Like Heaviside, Bentham often couldn't tolerate more than a single visitor at a time. Meeting people for the first time was agony. In order to gain some emotional control, Bentham found it necessary to orchestrate the precise time, location, and spatial position of the people during meetings. Even contacting people through the mail was difficult. For example, on several occasions Bentham would write fascinating letters to famous philosophers only to shrink from mailing them. Although his timidity prevented him from marrying a woman, he was not a homosexual. In fact, he did fall in love with a woman but was too shy to pursue her.

Bentham loved cats, pigs, and rodents. His favorites were his numerous trained rats that he stroked for hours. (In Appendix A, I'll tell you about a prominent woman whose best friends were rats.) Later he regretted spending so much time feeling fur between his fingers instead of being more profitably engaged in intellectual matters. He also had at least one trained pig that would follow him around like a dog.

Not only did Bentham substitute animals for human companions, but he also substituted household appliances for friends. For example, he gave

22. Jeremy Bentham's best friends.

human names to inanimate objects. He named his teapot "Dick." His walking stick was named "Dapple."

Bentham had a peculiar interest in the rituals of death. For example, to Bentham, cemeteries and burials were a waste of money. Instead, he suggested that embalmed corpses be mounted upright along stately drives and busy thoroughfares. I can just imagine his pleasure at seeing corpses planted like palm trees along the Santa Monica Boulevard or affixed to lampposts along New York's Fifth Avenue, for as far as his eye could see.

Bentham's will insisted that he be dissected and his body flayed open in front of his friends! Upon his death, his dreams of dissection were finally realized. After his organs were removed as his friends looked on in horror, his skeleton was padded with straw and dressed in a nice suit. His corpse was then set upright in a glass-fronted, mahogany case. The dissection was so comprehensive that his head had to be replaced with a wax copy, although his actual, mummified head was placed between his feet.

Should you ever be curious about visiting Bentham today, you can gaze upon his lifelike corpse and mummified head on display at the University College in London. His artificial eyes stare at you like Linda Blair's in *The Exorcist*. In his lifeless hand is his beloved walking stick, Dapple. His precious Dick is nowhere to be found.

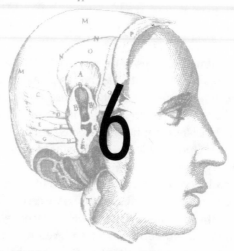

THE MUTTON MAN FROM LONDON

SINCE THE DEATH OF NEWTON, ENGLAND HAS
SUSTAINED NO SCIENTIFIC LOSS SO GREAT AS THAT
OF CAVENDISH.

SIR HUMPHRY DAVY (1778–1829)

 FACT FILE

- **Name:** Henry Cavendish
- **Born:** October 10, 1731, Nice, France
- **Died:** February 24, 1810, London, England
- **Occupation:** Physicist and chemist
- **Achievements:** Not only was Cavendish a gifted experimentalist, but he was also an expert mathematician. Cavendish determined the chemical composition of air and water, the nature and properties of hydrogen, the specific heat of various substances, and numerous properties of electricity. He measured the density and mass of the Earth in an ingenious experiment now known as the Cavendish experiment. He

was elected to the Royal Society, the Royal Society of Arts, and the Society of Antiquaries. He was a trustee of the British Museum, a manager of the Royal Institution, and a foreign associate of the Institut de France.

- **Marital status:** Celibate; never married
- **Religion:** Cavendish never identified himself with any religious body, and he never attended a church.
- **Appearance:** Cavendish was tall and gaunt, with a squeaky, hesitant voice. His clothes, 50 years behind the style, consisted of a faded crumpled violet suit with a high collar and frilled cuffs and a three-cornered hat.
- **Strange brain:** Cavendish exhibited excessive shyness (people were not permitted to look at him while speaking) and gynophobia (he found it unbearable to look at or talk to a woman) and was overly selective about foods (he ate only mutton).
- **Favorite quotes about Cavendish:**

> "Cavendish probably uttered fewer words in the course of his life than any man who ever lived to fourscore years, not at all excepting the monks of La Trappe." (Lord Brougham)

> "He did not love; he did not hate; he did not hope; he did not fear." (George Wilson)

☞ THE STRAIGHT DOPE

Four years after the death of Isaac Newton, the reclusive Henry Cavendish was born into one of England's wealthiest families. Figure 23 shows the Cavendish coat of arms with an inscription that translates to "Always be on your guard against danger."

Henry Cavendish's mother's health was so poor that she left England and traveled to sunny Nice, where Cavendish was born. She died after giving birth to his brother when Cavendish was only three years old.

During the years of 1749 through 1753 Cavendish attended Peterhouse College at the University of Cambridge, but he did not finish his degree, possibly because he refused to take the required oath to the Church of England. After a tour of Europe with his brother, he stayed with his father until his father's death in 1783. It was during this period of time that Cavendish conducted his famous electrical and chemical experiments.

Cavendish's father was a distinguished scientist and prominent figure in the counsels of the Royal Society, and he encouraged his son to pursue

23. The Cavendish family coat of arms. *Cavendo tutus* roughly translates to "Always be on your guard against danger."

science. In fact, Cavendish started his research as an assistant to his father. Although dad was a tightwad when it came to money, he did place all of his instruments at his son's disposal and introduced him into London's scientific circles.

For 30 years, Henry Cavendish lived on a very meager allowance from his father. However, around the time his father died, Henry Cavendish, age 40, became a virtual Donald Trump through inheritance from an aunt. According to Jean-Baptiste Biot, a contemporary French scientist, Cavendish became "the richest of all learned men, and very likely also the most learned of all the rich."

Cavendish's newly acquired wealth had virtually no effect on his living habits because he spent money on books and laboratory equipment but little else. He was in fact a shabby, eccentric man who rarely spoke. The few words he uttered were spoken hesitantly in a shrill, thin voice—like the squeaking of a mouse. When his banker called him to suggest an investment, Cavendish told him never to bother him again about his growing balance. "If it is any trouble to you," Cavendish told the banker, "I will take it out of your hands!"

Cavendish's brilliant science rarely achieved popular acclaim because he was mortified by public praise. However, he did accept the honor of being made a fellow of the Royal Society in 1760 and being elected one of the foreign associates of the Institut de France in 1803.

Over the years, Cavendish's research spanned the vast arenas of pure mathematics, mechanics, magnetism, optics, geology, and industrial science. Cavendish explained the molecular basis of heat long before the theory was publicized by his contemporaries, and he did much to improve the accuracy of the ordinary mercury thermometer. In chemistry, his main research was on the topic of gases. He recognized hydrogen as a separate substance and discovered that water is not an element but a compound of hydrogen and oxygen. He also discovered nitric acid. In his applied research, Cavendish designed methods for protecting gunpowder from lightning. He investigated the physical properties of gold alloys at the request of the government, which was concerned over the loss of gold in coins through wear.

Cavendish also conducted remarkable experiments with electricity, but since his work was largely unpublished, he received little fame. For example, he discovered that the potential across conductors was directly proportional to the current through them. He discovered that the force between a pair of electrical charges is inverse to the square of the distance between them. This law was subsequently established by Coulomb, a French physicist, and this fundamental principle is now known as Coulomb's law. Cavendish also developed the idea that all points on the surface of a good conductor are at the same potential with respect to a common reference, the Earth. This concept was important for further developments in electrical theory.

Much of Cavendish's "secret" experimental work was discovered, in his notebooks and manuscripts, nearly a century later by James Clerk Maxwell, a famous Scottish mathematical physicist mentioned in Chapter 2. For example, Figure 24 shows a page of Cavendish's notes on the electrical properties of heated glass. Many of Cavendish's electrical manuscripts are now published in *The Electrical Researches of Honorable Henry Cavendish*, edited by Maxwell (Cambridge, 1879). Some of Cavendish's lesser-known research includes work on the altitude of the aurora, a reconstruction of the Hindu civil year, a calculation on nautical astronomy, and a method of marking divisions on circular astronomical instruments.

In one of his most impressive experiments, Cavendish at the age of 70 "weighed" the world! In order to accomplish this feat, he didn't have to become an Atlas, but rather he determined the density of the Earth using highly sensitive balances illustrated in Figure 25. In particular, he used a torsional balance consisting of two lead balls on either end of a suspended beam. These mobile balls were attracted by a pair of stationary lead balls.

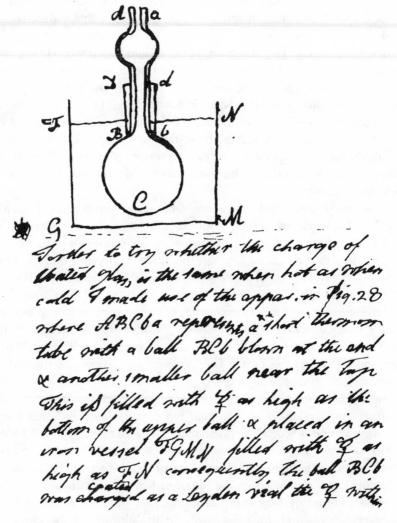

24. A page from Cavendish's notes on the electrical properties of heated glass.

To reduce air currents, he enclosed the device in a glass case and observed the motion of the balls from far away by means of a telescope. Cavendish calculated the force of attraction between the balls by observing the balance's oscillation period, and then computed the Earth's density from the force. He found that the Earth was 5.4 times as dense as water. Cavendish

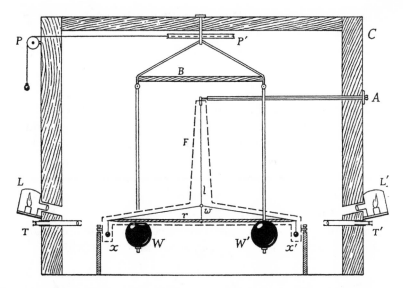

25. Apparatus used by Cavendish to determine the density of the Earth. The diameters of the attracting and attracted spheres are 12 and 2 inches, respectively.

was the first scientist able to detect minute gravitational forces between small objects. (The attractions were 1/500,000,000 times as great as the weight of the bodies.) By quantifying Newton's law of gravitation, he made the most important addition to gravitational science since Newton.

In many ways, Cavendish was like Isaac Newton. Both men hated criticism. However, the only time Cavendish actually replied to a critic was when Kirwan, the scientist discussed in Chapter 4, criticized some of his celebrated experiments on the composition of air. In an apologetic tone, Cavendish timidly begs forgiveness from the Royal Society for even addressing the conflict:

> As I do not like troubling the Society with controversy, I shall take no notice of the argument used by him, but shall leave them for the reader to form this judgement of; much less will I endeavor to point out any inconsistences or false reasonings, should any such have crept into it; but as there are two or three experiments mentioned there, which may perhaps be considered as disagreeing with my opinion, I beg leave to say a few words concerning them.

Cavendish then proceeds to demolish Kirwan's arguments.

Cavendish died at 78 after a few day's illness, and was buried in what is now Derby cathedral, England. Sir Humphry Davy (1778–1829) gave a moving eulogy:

> It may be said of him, what can, perhaps, hardly be said of any other person, that whatever he has done has been perfect at the moment of its production. Since the death of Newton, England has sustained no scientific loss so great as that of Cavendish. Like his great predecessor, he died full of years and of glory. His name will be an object of more veneration in future ages than at the present moment. Though it was unknown in the busy scenes of life, or in the popular discussions of the day, it will remain illustrious in the annals of science, which are imperishable as that nature to which they belong, and it will be an immortal honor to his age, and to his country.

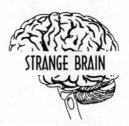

STRANGE BRAIN

Henry Cavendish was the greatest of all 18th-century scientists, and one of the greatest scientists who ever lived. Yet he wrote no books and fewer than 20 articles during his scientific research spanning fifty years. Not only is he one of the Britain' greatest scientists, but he is also the shyest—a distinction that made the vast extent of his scientific writings secret until after his death, and therefore his important discoveries are now associated with the names of subsequent researchers. This reminds me of the posthumous discovery of Heaviside's scientific papers that were crammed into his attic and used as insulation. The huge number of manuscripts uncovered after Cavendish's death show that he conducted extensive research in literally all branches of physical sciences of his day.

Cavendish rarely appeared or talked to people in public. Social interaction of any kind terrified him. The only people to whom he uttered a few words were scientists, but even at Royal Society gatherings he hardly ever spoke.

He was so shy around women that he only wrote notes to his house-keeper. Each night he would leave a scrap of paper on the hall table specifying his rather boring menu. I'm not sure why he even bothered, because he ate only the flesh of sheep. Why is it that so many of the geniuses in this book preferred uncomplicated meals and no women? The only known record of Cavendish speaking to his housekeeper is in regard to a scheduled dinner with several colleagues. Cavendish's housekeeper came to consult him regarding what should be ordered for dinner. Caven-dish's only reply was, "A leg of mutton." "But sir," she replied, "that will not be enough for five," to which he replied: "Well, then, get two."

After this incident, he ordered all his female housekeepers to keep out of sight. If he caught the slightest glimpse of them—an arm, a leg, or worse, a face—he would fire them on the spot. The one time he did see a female servant, he was so mortified that he built a second staircase for the servants' use so that he could avoid them.

The only known incident in which Cavendish acted chivalrously to a lady was in Clapham, where a woman was being chased by a mad cow. Cavendish was taking a solitary walk, saw the lady in distress, and stood with his body between the cow and the woman to prevent her from being attacked.

Although Cavendish was quite thrifty with himself and wore the shabbiest of clothes, he was generous with charities. Whenever invited to donate, he would ask the value of the largest previous gift and then match it. Some charities caught wind of this, and fabricated outrageously high prior donations.

Not only did Cavendish never marry, but he never formed any kind of friendship with a woman nor any member outside his family. Anyone he met was instructed to look away from him while talking. When someone violated this rule, Cavendish ran away, jumped into a cab, and sped home. Unlike Jeremy Bentham, whose shyness prevented him from pursuing a woman he loved, Cavendish appears never to have loved.

Cavendish was such a loner that when he was about to die, he ordered everyone out of the room. He wanted to die totally alone. He told his servants that he had "something to think about" and didn't want to be disturbed. After they quietly left, he died in total solitude at the age of 78.

After Cavendish's death, his valet searched Cavendish's chest of drawers and found parts of embroidered dresses and valuable jewelry including a lady's stomacher, richly decorated with diamonds. Soon it was discovered that Cavendish was the largest holder of bank shares in Britain, yet he never cared for the conveniences of wealth. His fortune had grown over the years to about 750,000 pounds. Upon Cavendish's death, his nephew George inherited most of Cavendish's money. It wasn't that Cavendish had been very close to George (Cavendish only permitted George to see him for exactly 0.5 hours each year), but there was no one else to whom he felt closer. An additional 15,000 pounds were left to Sir Charles Blagden, secretary of the Royal Society. Apparently Cavendish secretly liked his brief conversations with Blagden.

The fact that Cavendish left his wealth to virtual strangers is not unexpected considering Cavendish never seemed to love nor feel close to anyone. Much of the time he appeared to be more an alien or robot than a member of the human race. Would Cavendish have made fewer scientific discoveries if he could have taken a drug to increase his love and decrease his shyness? One wonders to what extent treatment with modern drugs such as Anafranil or Prozac, discussed in Chapter 10, would have changed Bentham and Cavendish, and thereby affected the course of history.

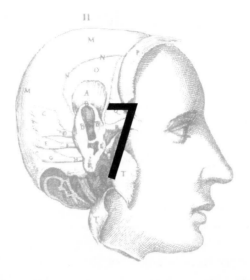

THE SPRAINED BRAIN FROM BIRMINGHAM

QUANTIFICATION WAS GALTON'S GOD. NO MAN
EXPRESSED HIS ERA'S FASCINATION WITH NUMBERS
SO WELL AS DARWIN'S CELEBRATED COUSIN,
FRANCIS GALTON.

STEPHEN J. GOULD

THOSE WHO HAVE NOT SUFFERED FROM MENTAL
BREAKDOWN CAN HARDLY REALIZE THE INCAPACITY
IT CAUSES, OR, WHEN THE WORST IS PAST, THE
CLOSENESS OF THE ANALOGY BETWEEN A SPRAINED
BRAIN AND A SPRAINED JOINT.

FRANCIS GALTON

THE RELIGIOUS INSTRUCTOR IN EVERY CREED IS
ONE WHO MAKES IT HIS PROFESSION TO SATURATE
HIS PUPILS WITH PREJUDICE.

FRANCIS GALTON

📁 FACT FILE

- **Name:** Francis Galton
- **Born:** February, 16, 1822, Birmingham, England
- **Died:** January 17, 1911, Haslemere, Surrey, England
- **Occupation:** Statistician, experimental psychologist, explorer, anthropologist, inventor, and eugenicist
- **Accomplishments:** A prolific author of over 300 publications, Galton achieved his greatest fame with his book *Hereditary Genius*, where he argued that talent (and virtually every human trait) was often inherited. He also conducted pioneering studies of human intelligence and statistics and was knighted in 1909. He coined the word "anticyclone" (meteorology) and "eugenics" (behavioral genetics). His research covered such diverse subjects as fingerprinting for personal identification, correlational calculus (a branch of applied statistics), blood transfusions, twins, criminality, and the art of exploration in undeveloped countries. Galton was a practical scientist, always testing his theories using apparatuses of his own design.
- **Little-known accomplishment:** Galton was the first to capture and bring strange, blind, Yugoslavian amphibians (Proteus) to England.
- **Appearance:** Galton had "intelligent, rather eager blue eyes and heavy brows, a long straight mouth, and bald head." (C. Markham, president of the Geographical Society)
- **Religion:** Galton was born to Quakers but had little interest in religion. He wrote to Charles Darwin that traditional biblical arguments made him "wretched." Later he suggested that the civilized world should give up all belief in the efficacy of prayer, and he became an opponent of Christianity.
- **Some Galton books:** *Hereditary Genius* (1869), *Inquiries into Human Faculty* (1883), *Tropical South Africa* (1853), *The Art of Travel; or, Shifts and Contrivances Available in Wild Countries* (1855), *Art of Campaigning* (1855), *Knapsack Guide for Travellers in Switzerland* (1864), *Decipherment of Blurred Finger Prints* (1893), *Probability, the Foundation of Eugenics* (1907), and *Memories of My Life* (1908)
- **Favorite book chapter title:** "Revolting Food that May Save the Lives of Starving Men," in *The Art of Travel*
- **Favorite paper titles:** "Arithmetic by Smell" (*Psychological Review*, 1884), "Intelligible Signals between Neighboring Stars" (*Fortnightly Review*,

1896), "Gregariousness in Cattle and in Men" (*Macmillan's Magazine*, 1861), and "Statistical Inquiries into the Efficacy of Prayer" (*Fortnightly Review*, 1872)

- **Favorite invention:** Galton invented a ventilating top hat with a movable lid that could be raised by squeezing a rubber bulb. He believed this would allow his overheated cranium to cool. He admitted the strangeness of his invention but said in defense that it was better than "falling into a fit upon the floor."

- **Strange brain:** Galton had an obsessive predilection for quantifying anything that he viewed—from the curves of women's bodies to the number of brush strokes used to paint his portrait. In 1897, he published a paper in *Nature* on the length of rope necessary for breaking a criminal's neck without decapitating the head. In short, Galton was obsessed with the idea that anything could be counted, correlated, and understood as some sort of pattern. Clements Markham (president of the Geographical Society) remarked "his mind was mathematical and statistical with little or no imagination."

- **Honors:** Galton was awarded the Royal Geographical Society's gold medal and elected to that body's council and to the Athenaeum Club. The French Geographical Society awarded him their Silver Medal, and he was made a Fellow of the Royal Society in 1856. He was awarded the Copley medal of the Royal Society in 1910.

- **Marital status:** Married, but his attachment to his wife does not appear to have been a romantic or sexual one. No children.

- **Galton's racist diary entry on managing African servants:** "Flog . . . do not hit a [expletive deleted] head. If wrestling, press at his spine if taken by throat . . . Boiling water, hot sand on their naked bodies."

☞ THE STRAIGHT DOPE

IF YOU HAVE HEALTH, A GREAT CRAVING FOR ADVENTURE, AT LEAST A MODERATE FORTUNE, AND CAN SET YOUR HEART ON A DEFINITE OBJECT, WHICH OLD TRAVELLERS DO NOT THINK IMPRACTICABLE—TRAVEL BY ALL MEANS.

FRANCIS GALTON
THE ART OF TRAVEL; OR, SHIFTS AND CONTRIVANCES AVAILABLE IN WILD COUNTRIES

Francis Galton, cousin of Charles Darwin, was born to a family of bankers and gunsmiths of the Quaker faith. His family life was happy. Galton's mother Violetta lived to 91, and most of her children lived to their nineties or near nineties. Perhaps the longevity of his ancestors accounted for Galton's very long life.

Before Galton was born, one of Violetta's children died, convincing Violetta to leave the Quaker sect and join the Church of England. As a result, Galton never knew the Quaker faith.

When Francis Galton was born, 12-year-old sister Adele asked to be his primary caretaker. She placed Galton's cot in her room and began teaching him letters, which he could point to and recognize before he could speak. He would cry if the letters were removed from sight.

By the age of two and a half, Galton could read a little book, *Cobwebs to Catch Flies*, and a month later he signed his own name. Some of his writing, produced when he was four years old, still survives, for example:

> The Lord is gracious and full of compassion, slow to anger and
> of great mercy. The Lord is good to all and his tender mercies
> are over all his works. Papa why do you call my books dirty,
> that one came from the Ware-House, I think they are very clean.

Soon after he wrote this, Adele caught Galton hoarding pennies. When she asked him why, he responded that he was saving money to help cover his eventual university admittance.

Toward the end of his fourth year, Galton's mental abilities exploded. Here is a letter to his sister:

> My dear Adele,
>
> I am four years old and can read any English book. I can say all the Latin substantives and adjectives and active verbs besides 52 lines of Latin poetry. I can cast up any sum in addition and multiply by 2, 3, 4, 5, 6, 7, 8, 9, 10, 11. I can also say the pence table. I read French a little and I know the Clock.
>
> Francis Galton, February-15-1827

When Galton was five years old, Adele sent him to the local school. Galton predicted his mother would remove him from school because the boys were so "vulgar"—having never heard of the *Iliad* or *Marmion*—but he remained in the school for three years, having been made the head boy on his arrival. By the time he left the local school at the age of eight and a half, he had read Ovid's *Metamorphoses* and *Epistles*, Eton's *Latin Grammar*, *Delectus*, Eutropius, and Phaedrus' *Fables*.

When Galton was seven, he became fascinated with insect classification. Here is an excerpt from his father's diary dated April 4th, 1830:

> I read him an extract about locusts in Peru, and Violetta said she'd seen lots at Ramsgate. Francis: "Oh, those were cockchafers." "Well," said Violetta, "they are the same thing." Francis: "Oh no, they're quite different, for the cockchafer belongs to the order Coleoptera but the locust belongs to the order Neuroptera."

Various psychologists have attempted to estimate Galton's IQ by comparing his early achievements to those of normal children of different ages. For example, Galton's clock and multiplication skills at age four are comparable to skills normally attained by other children between the ages eight and nine. Therefore, Galton's mental age was double his chronological age, and, according to the Binet system of IQ measurement where IQ = (mental age/chronological age) times 100, Galton's IQ was therefore 200. Although IQs today are expressed in slightly different terms of deviation from a mean score of 100, there's no doubt that Galton's IQ was incredibly high.

When Galton was 14, he was sent to a high school that he considered inferior. He writes in his autobiography:

> The character of the education was altogether uncongenial to my temperament. I learnt nothing, and chafed at my limitations. I had craved for what was denied, namely, an abundance, of good English reading, well-taught mathematics, and solid science. Grammar and the dry rudiments of Latin and Greek were abhorrent to me, for there seemed so little sense in them.

He considered schooling as, "that unhappy system of education that has hitherto prevailed, by which boys acquired a very imperfect knowledge of the structure of two dead languages, and none at all of the structure of the living world." (Today, psychologist Leta Hollings would agree with Galton. She has said that the student of IQ 140 wastes half his time in the ordinary classroom, while a student with IQ 180 wastes almost all of it.)

Galton's parents dreamed their son would become a medical doctor, and Galton spent a great deal of time in his teenage years exploring Europe in search of good schools. He trained at hospitals in London and Birmingham but then decided to interrupt his medical education to pursue a Cambridge degree in mathematics.

JOURNEYS

In 1840, Galton was bored by life in England and felt the urge to explore the world. "A passion for travel seized me," he said, "as if I had been a migratory bird." For the next decade, he would embark on a shattering odyssey of self-discovery. In fact his biography reads more like Pirsig's *Zen and the Art of Motorcycle Maintenance* or Simon's *Jupiter's Travels* than a Victorian genius'. Galton suddenly moved like a roller-coaster over some of the world's most mysterious physical and psychological terrain: raids on the Abyssinian border, camel rides through Egyptian desserts, death-defying escapes in the jungles of South Africa. . . . I'll give only a brief overview of his adventures, but you should consult his original writings to be enthralled by Galton's own descriptions of the exotic places and people, by his ability to adjust to adversity, by his humor and incisiveness, but above all by the realization that to understand his world, he had to make himself vulnerable to it so that it could change him.

Egypt and Syria

It is time to leave England and explore! Galton's journey began at the University of Giessen, Germany, where he started attending chemistry lectures. Still craving adventure—never having patience to complete any of his studies, and before his father could say anything to the contrary— Galton left Giessen by coach to Passau on the Danube. Unfortunately Passau's Danube steamer was under repair, so he took a small boat with one oarsman to travel down to Vienna. Next he caught a steamer to Budapest and then went to Constantinople where he wrote his father:

> Here I am at Constantinople among Turks, Armenians, Greeks, Jews and Franks in a good Lodging House, as well as possible, and happier and happier every day. . . . There are veiled ladies just looking out between folds of gauze and very pretty eyes they have too. . . . I saw the woman's slave-market today. If I had 50 pounds at my disposal I could have invested in an

excessively beautiful one, a Georgian. . . . Most of the black ones were fettered, but they seemed very happy dancing and singing and looking on complacently while a couple of Turks were wrangling about their prices.

Another boat took him to Smyrna and Syria, then on to Athens and Trieste.

While exploring the caves of Adelsberg, Yugoslavia, he discovered bizarre blind amphibians named Proteus, and he was determined to bring specimens back to Britain. He took a steamer to Venice and then sat day and night nursing the Proteus on the week's journey from Milan to Boulogne. While crossing the Alps, he kept Proteus in a bottle placed under his coat to prevent the water from freezing. Their gills and four short legs wiggled periodically as they sloshed in the dark liquid. One can only wonder how Galton's fellow passengers reacted to having foot-long, eel-like creatures as neighbors. Galton finally made it back to England, and achieved the distinction of being the first person ever to capture and bring the slippery Proteus to England.

After his episodes with the blind amphibians, Galton attended Trinity College, Cambridge, where he invented a "Gumption-Reviver" machine to invigorate him during protracted studies. The machine continuously wet his head with cold water during the difficult period of study from ten o'clock p.m. to two o'clock a.m.

In his third year, Galton had a nervous breakdown resulting from overwork, and had to return home. He compared his illness to a mill, turning inside his head, that wouldn't allow him to banish obsessive thoughts. He felt pain when looking at a printed page. Luckily, he recovered quickly after moderating his intense lifestyle.

While at Cambridge he invented various devices: a new type of oil lamp, a lock, a balance, and a rotating vane steam engine. He became famous when he entered a lion's den at a traveling circus visiting Cambridge. Apparently, he was only the fourth person to have attempted such a stunt.

Galton left Cambridge without a degree and continued medical studies in London. Perhaps he would have finished his medical degree if his father hadn't died when Galton was 22 years old. His father left him a "sufficient fortune" to make him "independent of the medical profession." Therefore Galton promptly dropped out of medical school.

With his newly acquired money, Galton became a world explorer. He left London in 1845 and journeyed to France and on to Malta. On a steamer from Malta to Alexandria he met two Cambridge friends. Together they reached Cairo and sailed along the Nile.

At one point on his journey, Galton found the Nile filled with hippopotamuses, and he brought out a rifle and began to fire at them. Lucky for the hippos, Galton was a poor shot. Despite a constant expenditure of ammunition, he didn't hit a single hippo. After sailing to Khartoum and to Metemmneh, Galton hired camels to traverse the Bayouda Desert to Dongola. Galton also bought two monkeys. Unlike this success with the Proteus, the monkeys did not survive more than a night in England, where they froze to death in one another's arms.

From Dongola, Galton went to Haifa, Cairo, and Damascus, where he took lessons in Hebrew and watched many Jewish ceremonies. He also became fluent in Arabic. He grew interested in the Arabs and their religion. Islam appealed to Galton as much as Christianity ever had. Next he visited Aden in Lebanon, then Tripoli, Beirut, Baghdad, Jerusalem, Marseilles, Paris, and finally, in 1846, went back to London.

A few years later, Galton spent the entire summer in the Shetlands shooting at seals and descending dangerous cliffs. He left the Shetlands in 1849 with a crate crammed with an assortment of live seabirds for his brother's estate. Again, his success was not as great as with Proteus because three-quarters of the birds died of cold on the railway journey south. The remainder thrived for a while at a relative's freshwater lake.

Galton began to brood over his own future and lack of goals, but for four more years he continued his free-wheeling explorations and sporting trips. There was no real scientific purpose to his ramblings, but the survival skills he learned would prove invaluable in an important future exploration.

Southwest Africa

In 1849, Europe knew very little about Africa. The entire southern interior was an empty space on the map, and even the courses of large rivers were uncharted. In 1850, Galton set sail for Cape Town on a small boat. Large winds swept him as far west as South America, and as a result, it required 86 days to reach Cape Town. While traveling, Galton occupied his time by learning the Bechuana language at a rate of 20 words a day.

The Sprained Brain from Birmingham

He also practiced with a sextant, a device for measuring angular distance between objects and for observing altitudes of celestial objects to determine latitude at sea.

When he arrived in Africa, Galton hired seven servants in Cape Town, sailed to Walfisch Bay, and prepared to enter Damaraland, which had never been explored by Europeans. To make his South African journey easier, he brought many presents for the native people, including guns, knives, tobacco in 1 oz. sticks, beads, chains, ornamental belts, bright uniforms, Jew's harps, and a cheap theatrical crown that he vowed "to place on the head of the greatest or most distant potentate I should meet with in Africa." As Galton traveled inland to Otjimbingue, one of his horses was killed by a mule. Galton cut some meat off the dead animal and stored it for later consumption.

Next he went on to Barmen, where he said he saw some horrible sights:

> I saw two poor women, one with both legs cut off at her ankle joints, and the other at one. They had crawled the whole way on that eventful night from Schelen's Hope to Barmen, some twenty miles. The Hottentots had cut them off after their usual habit, in order to slip off the solid iron anklets that they wear. These wretched creatures showed me how they had stopped the blood by poking the wound stumps into the sand. . . . One of [the tribal leader's evil] sons, a hopeful youth, came to a child that had been dropped on the ground and who lay screaming there, and he gouged out its eyes with a small stick.

While waiting for permission to pass through the tribal leader's land, Galton spent a month at Otjimbingue learning the Damara language. There he came upon a voluptuous, beautiful Hottentot woman, who became the object of his obsession for measurements. (Hottentots are a race of Southwest African people.) Galton immediately began to assess her attractive figure using his sextant:

> I profess to be a scientific man, and was exceedingly anxious to obtain accurate measurements of her shape; but there was difficulty in doing this. I did not know a word of Hottentot, and could never therefore have explained to the lady what the object of my footrule could be; and I really dared not ask my worthy missionary host to interpret for me. The object of my admiration stood under a tree, and was turning herself about to

all points of the compass, as ladies who wish to be admired usually do. Suddenly my eye fell upon my sextant; the bright thought struck me, and I took a series of observations upon her figure in every direction, up and down, crossways, diagonally, and so forth, and I registered them carefully upon an outline drawing for fear of any mistake: this being done, I boldly pulled out my measuring tape, and measured the distance from where I was to the place where she stood, and having thus obtained both base and ankles, I worked out the results by trigonometry and logarithms.

(One wonders if Galton would be equally analytical when viewing a modern copy of *Playboy*. Perhaps the fact that the woman was black made her more of an object to him—to be measured—despite his professed admiration.)

The next stop was Lake Omanbonde. To help him on his journey, Galton employed the services of Damara men and women. As they travelled, Galton became angry at some of their food taboos which made it harder to feed them. He writes, "There is hardly a particle of romance, or affection, or poetry, in their character or creed; but they are a greedy, heartless, silly set of savages." Galton even boasted of whipping them at times.[1]

Galton was frustrated by the inability of his Damara guides to tell him the length of time required to travel to Lake Omanbonde. Part of the problem was due to the Damara's never using numerals greater than three. Galton commented on the Damara's numerical limitations:

When they wish to express four, they take to their fingers, which are to them as formidable instruments of calculations as a sliding-rule is to an English schoolboy. They puzzle very much after five; because no spare hand remains to grasp and secure the fingers that are required for "units." Yet they seldom lose oxen: the way in which they discover the loss of one is not by the number of the herd being diminished, but by the absence of a face they know.

Galton discovered that the Damara had an amazing number of descriptive names for cattle, with every imaginable variation in color patterns nameable. In fact, the Damara used more than 1000 words to describe the different markings and colors of animals.

After Galton reached the dried lake bed of Lake Omanbode, he forged on ahead to the land of the Ovampo. There his sexual insusceptibility offended an Ovampo king who hospitably presented him with the Princess Chipanga as a temporary wife. Galton wrote:

> I found her installed in my tent in Negress finery, riddled with red ochre and butter, and as capable of leaving a mark on anything she touched as a well-inked printer's roller. I was dressed in my one swell-preserved suit of white linen, so I had her ejected with scant ceremony.

After many adventures, Galton finally returned to London where the Royal Geographical Society gave him a gold medal for his incredible journey. The citation was awarded

> for having at his own cost and in furtherance of the expressed desire of the Society, fitted out an expedition to explore the center of South Africa, and for having so successfully conducted it through the countries of the Namaquas, the Damaras, and the Ovampo (a journey of about 1700 miles), as to enable this Society to publish a valuable memoir and map in the last volume of the Journal, relating to a country hitherto unknown; the astronomical observations determining the latitude and longitude of places having been most accurately made by himself.

Soon after this Galton was heaped with honors for his explorations. He was elected to the Council of the Royal Geographical Society and to the Athenaeum Club. The French Geographical Society awarded him their Silver Medal, and he was made a Fellow of the Royal Society in 1856. A full account of his expedition is described in his 1853 book *Tropical South Africa*. Even the famous Charles Darwin wrote to Galton to congratulate him on the fascinating and lively book:

> I last night finished your volume with such lively interest that I cannot resist the temptation of expressing my admiration at your expedition, and at the capital account you have published of it. . . . What labors and dangers you have gone through: I can hardly fancy how you can have survived them, for you did not formerly look very strong, but you must be as tough as one of your own African waggons! I live at a village called Down near Farnbourough in Kent, and employ myself in Zoology; but the

objects of my study are very small fry, and to a man accustomed to rhinoceroses and lions, would appear infinitely insignificant.

MARRIAGE AND THE ART OF TRAVEL

Around 1853 Galton seems to have gotten the wanderlust out of his system. He married Louisa Bulter and settled into a peaceful London home where he stayed until his death over a half century later. Most of his experiments were done at home where he also wrote nine books and some 200 papers. For the rest of his life, his only trips away from home were occasional European vacations. The newly wed Galtons had a honeymoon tour starting in Switzerland, which became their favorite choice for subsequent holidays.

Galton was always quite kind and unselfish toward Louisa, although his interest in her did not seem to have been a very romantic or sexual one. However, Louisa was quite intelligent and shared in many of her husband's interests. Later when Galton refuted Christianity, he never let his disbelief dominate Louisa, and he attended church and conducted prayers for the household without protesting.

Galton did not have children, and at times he seemed to be asexual. His biographer D. W. Forrest writes, "There is no trace of heterosexual interest after 1846, apart from his marriage in 1853, and for the remainder of his life his attitude to women is one of polite indifference, so much so that Darlington has referred to him as a 'natural celibate.'"

Galton's home with Louisa was a monument to practicality, and his odd assortment of furniture was an interior decorator's nightmare. For example, he equipped his bathroom at the top of the stairs with a rod. When the bathroom was occupied, and the bolt slid in place, the rod could be clearly seen from the bottom of the stairs. Thus, a person would not have to waste time climbing the stairs only to disturb the person on the toilet.

Galton didn't read very much when compared to some of the other geniuses in this book. In fact, his personal library was quite small, consisting mostly of autographed copies of books that colleagues sent him.

Galton's successful 1855 book *The Art of Travel; or, Shifts and Contrivances Available in Wild Countries* overflowed with advice for survival on expeditions. Here are some notable, sometimes racist, passages from his travel book:

The Sprained Brain from Birmingham

- *Psychological Advice*: "Interest yourself chiefly in the progress of your journey, and do not look forward to its end with eagerness. It is better to think of a return to civilization, not as an end to hardship and a haven from ill, but as a close to an adventurous and pleasant life."
- *Medical Advice*: "The traveller who is sick, away from help, may console himself with the proverb that though there is a great difference between a good physician and a bad one, there is very little between a good one and none at all."
- *Bones*: "Bones contain a great deal of nourishment, which is got at by boiling them, pounding their ends between two stones, and sucking them. There is a revolting account in French history of a besieged garrison of Sancerre digging up the graveyards for bones as sustenance."
- *Living Flesh*: "The truth of Bruce's well known talk of the Abyssinians and others occasionally slicing out a piece of a live ox for food is sufficiently confirmed. . . . When I travelled in South-West Africa, at one part of my journey a plague of bush-ticks attacked the roots of my oxen's tails. Their bites made festering sores, which ended in some of the tails dropping bodily off. . . . The animals did not travel the worse for it. Now oxtail soup is proverbially nutritious."
- *Hiding Your Jewels*: "Before going to a rich but imperfectly civilized country, travelers sometimes buy jewels and bury them in their flesh. They make a gash, put the jewels in, and allow the flesh to grow over them as it would over a bullet. . . . A traveller who was thus provided would always have a small capital to fall back upon, though robbed of everything he wore."
- *To Make Your Savages Happy*: "Interrupt the monotony of travel by marked days on which you give extra tobacco and sugar to the servants. Recollect that a savage cannot endure the steady labor that we Anglo-Saxons have been bred to support. His nature is adapted to alternations of laziness and of severe exertion. Promote merriment, singing, fiddling, and so forth, with all your power."
- *On Women*: "It is the nature of women to be fond of carrying weights. You may see them in omnibuses and carriages, always preferring to hold their baskets or their babies on their knees to setting them down on the seats by their sides. A woman whose modern dress includes, I know not how many cubic feet of space, has hardly ever pockets of sufficient size to carry small articles; for she prefers to load her hands with a bag or other weighty object."

Any religious beliefs that Galton had did not survive the 1859 publication of Darwin's *Origin of Species*. In his own words, the book abolished "the constraint of my old superstition as if it had been a nightmare." Galton subsequently removed all references to Adam and Eve in his descriptions of the origin of fire in his 1867 edition of *The Art of Travel*. In his autobiography, Galton wrote:

> The publication in 1859 of the *Origin of Species* by Charles Darwin made a marked epoch in my own mental development, as it did in that of human thought generally. Its effect was to demolish a multitude of dogmatic barriers by a single stroke, and to arouse a spirit of rebellion against all ancient authorities whose positive and unauthenticated statements were contradicted by modern science.

In 1866, Galton's obsession with measurement may have been partly responsible for a second nervous breakdown. (The reasons for his breakdowns are not clear.) Galton wrote that outdoor exercises and travel were the best antidotes for his breakdowns:

> The warning I received in 1866 was more emphatic and alarming than previously, and made a revision of my mode of life a matter of primary importance. Those who have not suffered from mental breakdown can hardly realize the incapacity it causes, or, when the worst is past, the closeness of the analogy between a sprained brain and a sprained joint. In both cases, after recovery seems to others to be complete, there remains for a long time an impossibility of performing certain minor actions without pain and serious mischief, mental in the one and bodily in the other. This was a frequent experience with me respecting small problems, which successively obsessed me day and night, as I tried in vain to think them out. These affected mere twigs, so to speak, of the mental processes, but for all that most painfully.

RESEARCH AND SCIENCE

Galton pursued his research with the same vigor he did his world travels. Among Galton's earliest research areas was meteorology, where he studied anticyclones. In statistics, he originated the ideas of regression and correlation. In psychology, his ideas lead to mental testing and the measuring of the senses. Galton's most famous work, *Hereditary Genius* (1869), discussed evidence for the inheritance of scholarly, athletic, and artistic talent. He wrote:

> I have no patience with the hypotheses occasionally expressed, and often implied, especially in tales written to teach children to be good, that babies are born pretty much alike, and that the sole agencies in creating differences between boy and boy, and man and man, are steady application and moral effort. It is in the most unqualified manner that I object to pretensions of natural equality. The experiences of the nursery, the school, the university, and of professional careers, are a chain of proofs to the contrary.

Galton divided the range of above-average intellectual abilities into eight classes: A, B, C, D, E, F, and G (highest ability). He used lowercase letters to denote eight classes below the average: a, b, c, d, e, f, and g (lowest ability). He wrote:

> Eminently gifted men are raised as much as above mediocrity as idiots are depressed below it; a fact that is calculated to considerably enlarge our ideas of the enormous difference of intellectual gifts between man and man. I presume the class F of dogs and other of the more intelligent sort of animals, is nearly commensurate with the f of the human race, in respect to memory and powers of reason. Certainly the class G of such animals is superior to the g of humankind.

Galton believed that those people of highest intellect (class G) should require no schooling. In fact, he considered G intellects virtually independent of ordinary school education and better off educating themselves:

> People are too apt to complain of their imperfect education, insinuating that they would have done great things if they had been more fortunately circumstanced in youth. But if their power of learning is materially diminished by the time they have rediscovered their want of knowledge, it is very probable that their abilities are not of a very high description, and that, however well they might have been educated, they would have succeeded but little better.

Galton believed that religious people give birth to children exhibiting behavioral extremes; that is, the children are either pious or very immoral. He believed that religious people have strong moral tendencies but are highly unstable. For example, at one time they may appear selfish and sensual, and another altruistic. Galton writes, "The amplitude of the moral oscillations of religious men is greater than that of others whose *average* moral position is the same."

Galton, in his racist way, argued that black people rarely show intelligence as high as exhibited by class *F* white people, and that the number of mentally deficient blacks is very large. "The mistakes the Negroes made in their own matters," Galton wrote, "were so childish, stupid, and simpleton-like, as frequently to make me ashamed of my own species." Galton believed that black people were two grades below whites of his day and that the Athenians of 530–430 B.C. were two grades above whites of his day:

> This estimate, which may seem prodigious to some, is confirmed by the quick intelligence and high culture of the Athenian commonality, before whom literary works were recited, and works of art exhibited, of a far more severe character than could possibly be appreciated by the average of our race, the caliber of whose intellect is easily gauged by a glance of the contents of a railway bookstall.

(What would Galton have thought of a typical "bookstall" in Grand Central Terminal today? Using Galton's reasoning, shouldn't Greeks be to this day the most intelligent group, since they would have passed down this brilliance?)

Galton disagreed with church policies, particularly when it came to imposing celibacy on priests and nuns, because this biologically selected for aggressive, ignorant people in favor of gentle, intelligent people:

> Whenever a man or woman was possessed of a gentle nature that fitted him or her to deeds of charity, to mediation, to literature, or to art—the social condition of the time was such that they had no refuge elsewhere than in the bosom of the Church. But the Church chose to preach and exact celibacy. The consequence was that these gentle natures had no continuance, and thus, by a policy so singularly unwise and suicidal that I am hardly able to speak of it without impatience, the Church brutalized the breed of our forefathers.

Galton also commented on the unfortunate persecution of intellectuals by the Church and the rampant breeding of the unintelligent:

> Those she reserved on these occasions, to breed the generations of the future, were the servile, the indifferent, and again, the

stupid. . . . In consequence of this inbred imperfection of our natures, in respect to the conditions under which we have to live, we are, even now, almost as much harassed by the sense of moral incapacity and sin, as were the early converts from barbarism, and we steep ourselves in half-unconscious self-deception and hypocrisy, as a partial refuge from its insistence. Our avowed creed remains at variance with our real rules of conduct, and we lead a dual life of barren religious sentimental-ism and gross materialistic habitudes.

The publication of *Hereditary Genius*, containing these and more obser-vations, was criticized by nonscientific reviewers, primarily because of its harsh stance against the church. However, *scientists* of his day found it of great interest. For example, Charles Darwin wrote:

My Dear Galton,

I have only read about 50 pages of your book (to Judges), but I must exhale myself else something will go wrong in my inside. I do not think I ever in all my life read anything more interesting and original—and how well and clearly you put every point. . . . You have made a convert of an opponent in one sense, for I have always maintained that excepting fools, men did not differ much in intellect, only in zeal and hard work; and I still think this is an eminently important difference. I congrat-ulate you on producing what I am convinced will prove a memorable work. I look forward with intense interest to each reading, but it sets me thinking so much that I find it very hard work; but this is wholly the fault of my brain and not your beautifully clear style.

Yours most sincerely,
Charles Darwin

HUMAN FACULTY

In 1879, Galton became interested in how people use visual imagery to understand the world. In order to gain information, Galton handed ques-tionnaires on this subject to relatives, friends, and schools. One of his most interesting respondents was George Bidder, who had the mathematical

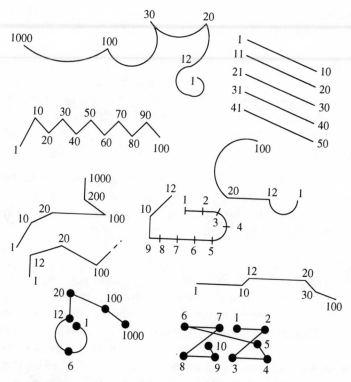

26. George Bidder, a human "calculator," visualized numbers in the form of spatial frameworks such as those shown in this illustration.

gifts of his father, a rapid calculator. When a number occurred to Bidder, he visualized it in the appropriate position on a spatial pattern made up by other numbers. Galton collected these instances of number forms in a *Nature* paper titled "Visualized Numerals" (Figure 26).

In 1883, Galton published his book *Inquiries into Human Faculty and Its Development*. In it are several hundred number forms, including examples from 22 families whose tendencies to visualize numbers appeared to be hereditary. Galton also found that some people associated the sounds of words with colors. For example, one of his subjects visualized vowels in the following manner:

- A—pure white, china-texture
- E—red, always opaque

- I—light bright yellow
- O—black, transparent, the color of deep water observed through thick clear ice
- U—purple

Recall that I discussed "synesthesia," or sensory cross-over, in Chapter 1 on Nikola Tesla.

Galton was also interested in extremes of human and animal hearing. To carry out his experiments, he created a device for producing extremely high notes near the human auditory limit. For animal experiments, Galton placed the whistle, now known as the "Galton whistle," in a hollow walking stick equipped with a bulb that he would press to inconspicuously make a sound. His experiments were reported in *Human Faculty*:

> I hold the stick as near as is safe to the ears of the animals, and when they are quite accustomed to its presence and heedless of it, I make it sound; then if they prick up their ears it shows that they hear the whistle; if they do not it is probably inaudible to them. Still, it is very possible that in some cases they hear but do not heed the sound.

On one expedition through Regent's Park Zoological Gardens, he annoyed the lions and did not get very many useful results.

Galton did find that cats have the most sensitive hearing as well as small dogs. Large dogs seemed to have less acute hearing. He commented:

> At Berne, where there appear to be more large dogs lying idly about the streets than in any other town in Europe, I have tried the whistle for hours together, on a great many large dogs, but could not find one that heard it. . . . I once frightened a pony with one of these whistles in the middle of a large field. My attempts on insect hearing have been failures.

Toward the end of *Human Faculty*, Galton talked about his own religious beliefs. "The religious instructor in every creed," he says, "is one who make it his profession to saturate his pupils with prejudice." In one particularly enigmatic passage, Galton talks about God and the fourth dimension:

> It is difficult to withstand a suspicion that the three dimensions of space and the fourth dimension of time may be four indepen-

dent variables of a system that is neither space nor time, but something else wholly unconceived by us. Our present enigma as to how a First Cause could itself have been brought into existence ... may be wholly due to our necessary mistranslation of the four or more variables of the universe, limited by inherent conditions, not the three unlimited variables of Space and Time.

Galton believed in the value of eugenics programs that fostered talent and healthiness and reduced stupidity and sickliness. Galton coined the word "eugenics" to denote plans to increase the proportion of people with good genetic endowments through selective mating of marriage partners. He believed that early marriage leads to more offspring and therefore should be encouraged in people with good hereditary qualities.

In 1884, Galton set up an "anthropometric laboratory" at the International Health Exhibition in South Kensington. Visitors would enter a passageway along which various measuring instruments were placed. For a cost of threepenny, subjects passed by, one at a time, while Galton and his staff measured their weight, height, arm span, breathing capacity, arm strength, reaction time, eyesight, hearing, and much more. Visitors received the results at the end of the assembly line, and Galton kept a copy for his own records.

Galton was very interested in human fingerprints, and his classification schemes used for personal identification are generally still in use today. To his disappointment, he found no correlations of fingerprint patterns with intelligence, morality, and race:

> I have prints of eminent thinkers and of eminent statesmen that can be matched by those of congenital idiots. No indication of temperament, character, or ability are to be found in finger marks, so far as I have been able to discover.

To increase the range of his sample, his colleagues sent him fingerprints from Welsh and blacks, and he took fingerprints from Basques and Jews. The only notable difference he found was that Jews had fewer arches and more whorls than non-Jewish English people.

In 1903, Galton wrote,

> There is no question that the pick of the British race are as capable human animals as the world can at present produce. Their defects lie chiefly in the graceful and sympathetic sides of

their nature, but they are strong in mind and body, truthful and purposive, and excellent leaders of the people of the lower races.

EXTRATERRESTRIALS, AND THE DEATH OF GALTON

Most people don't realize that Galton was also interested in the search for extraterrestrial intelligence. In his paper "Intelligible Signals between Neighboring Stars" (*Fortnightly Review*, 1896), he describes a signaling code that could be used to establish communication between us and Martians. Galton first assumed that any advanced civilization knows mathematics and would use it to convert images from a series of flashed signals. He suggested a ternary system using 3 symbols (dot, dash, line) that represent numbers. Examples of addition and multiplication would be first transmitted. The next message should be astronomical and convey the mean distance between the sun and the five planets, rotation time, and radii. Once the aliens understood what we meant by radius, the aliens would then be expected to send the value of pi and then send information on symbols for such formulas as πr^2 and $2\pi r$.

Galton also conducted research on prison sentence length, showing certain subconscious patterns in the sentences dealt by judges. In 10,000 sentences, he discovered high preferences for 3, 6, 9, 12, 15, 18, and 24 months with no prison terms of 17 months and very few at 11 or 13. Galton concluded in a *Nature* paper that subconscious bias for these term lengths was unjust because there was no correlation between the numbers and the severity of the crime. In 1897, he published another paper in *Nature* on the length of rope necessary for breaking a criminal's neck without decapitating the head. Galton found an error in previous force formulas that did not take into account the large neck muscles in fat men.

Galton's wife Louisa died at the age of 75. Galton wrote to his sister Emma:

> I had a nurse to sit up through the night who woke me at 2½ a.m. when dear Louisa was dying. She passes away so imperceptibly that I could not tell when, within several minutes. Dying is often easy! I cannot yet realize my loss. The sense of it will come only too distressfully soon, when I reach my desolate home. . . . Dear Louisa, she lies looking peaceful but worn, in the next room to where I am writing, with a door between. I

have much to be thankful for in having had her society and love for long. I know how you loved her and will sympathize with me.

(Oddly, Louisa's autopsy revealed that her stomach was less than one-third the normal size and the outlet was extremely narrow.)

In the winter of 1910, Galton's asthma became very bad. Before his death, Galton seemed to have realized that the end was near. He even joked that the hallway to his bedroom would be difficult to negotiate with his coffin. He soon became so weak that he could speak only a word or two, although his mind seemed clear. On January 17, 1911 he refused food, quoting Burns: "Some hae meat and canna eat, And some wad eat that want it."

Finally Galton was given oxygen as he struggled to ask his nephew to explain to the doctor that Galton had once done experiments with oxygen. An hour later, Francis Galton died at the age of 89.

STRANGE BRAIN

GALTON WAS VERY CLEVER AND PERFECTLY STRAIGHT IN ALL HIS DEALINGS WITH A STRONG SENSE OF DUTY. WITHOUT AN ATOM OF VANITY HE HELD TO HIS OWN OPINIONS AND AIMS TENACIOUSLY. HIS MIND WAS MATHEMATICAL AND STATISTICAL WITH LITTLE OR NO IMAGINATION. HE WAS NOT ADAPTED TO LEAD OR INFLUENCE MEN. HE COULD MAKE NO ALLOWANCE FOR THE FAILINGS OF OTHERS AND HAD NO TACT.

CLEMENT MARKHAMS, PRESIDENT OF THE GEOGRAPHICAL SOCIETY

Of all the "Strange Brains" in this book, Galton seems to be the least bizarre of the bunch. He didn't paint his nails pink, eat only mutton, or insist food come in multiples of three. However, his extreme intelligence, combined with his obsession for measuring, make him an ideal subject for this book. In this section, we'll get into his excessive passion for measurement.

As I mentioned in the Preface of this book, Galton, while working in a hospital, decided to sample every drug in the pharmacy in alphabetical order. His reasons are not clear, but we can speculate that his passion for testing got the better of him! Starting with the letter A, he began to take small doses. His peculiar quest came to an abrupt end with the letter C as a result of castor oil's laxative effects. (I'm glad they didn't have arsenic in the dispensary.)

Galton occasionally hallucinated when ill:

> When fancies gathered and I was on the borderland of delirium, I was aware of the imminence of a particular hallucination. There was no vivid visualization of it, but I felt that if I let myself go I should see in bold relief a muscular blood-stained crucified figure nailed against the wall of my bedroom opposite to my bed. What on earth made me think of this particular object I have no conception. There was nothing in it of the religious symbol, but just a prisoner freshly mauled and nailed up by a brutal Roman soldier. The interest in this to me was the severance between the state of hallucination and that of ordinary visualization. They seemed in this case to be quite unconnected.

Galton was extremely secretive about his personal affairs, and therefore we know only a little about his personal life. For example, even after his brothers' deaths he kept a file on the details of a quarrel he had with them. Galton used an indecipherable code to store the information.

Galton's wealth gave him the rare freedom to devote his considerable energy and intelligence to his favorite subject: numerical measurements. As you should surmise from the "Straight Dope" section, he had an obsessive urge to quantify everything he saw around him. If something were quantifiable, he could not resist getting out his sextant or ruler or making some statistical calculation. For example, one of his earliest papers dealt with a statistical test of the efficacy of prayer and the longevity of people who pray and who are prayed for. As evidence that prayers are not helpful, he calculated the average age at which people in various professions died:

Profession	Average age of death
Clergymen	66.42
Lawyers	66.51
Doctors	67.04

Since the pious did not live particularly long lives, Galton concluded, "The prayers of the clergy for protection against the perils and dangers of the night, for protection during the day, and for recovery from sickness, appear to be futile in result."

Galton also demonstrated that public prayer for kings, queens, and other leaders was completely ineffective because sovereigns were the shortest lived of all those who lived in relative wealth.

Position	Average age of death
Sovereign of the State	64.04
English Aristocracy	67.31
English Gentry	70.22

Missionaries who frequently prayed, and who were prayed for, often died in savage countries and did not live longer than totally immoral people.

Galton further assailed the notion that prayer helped the sick and dying by noting that if this were true, then insurance companies would have adjusted their policies for people who prayed. They did not, nor did they adjust policies for buildings, ships, or homes owned by religious and nonreligious people. (If God protected the homes of religious people more than nonreligious people, shouldn't insurance policies be less for religious people? If God equally protected the homes of priests and axe murders, what good then is prayer and piety?)

Galton noted that the placement of lightning protectors on churches was once seen to be a slap in the face of God, but people learned it was necessary to add devices to protect the church. He further reasoned that since so many rituals of religion were considered ineffective during his time (e.g., witch burning, laying on of hands by kings and queens), his contemporaries should equally abandon the idea that prayer has any effect. In fact, Galton claimed that the evidence is consistently negative.

Not all of Galton's contemporaries agreed with his arguments on prayer. For example, George Romanes, one of Galton's critics, suggested that statistical methods were ineffective for studying the efficacy of prayer. Ramanes did not know if prayers for longevity were answered, but he pointed out that only a proportion of the clergy prayed for a long life for themselves. If only one out of eight clergymen had their prayers answered, then the fact that some clergy appear to live two years longer than other groups may have concealed the fact that one of the eight lives 16 years longer.

Galton's sister Emma was unhappy about the section of his book on prayers. She commented:

> I cannot help greatly deploring what you have said on prayer. Whatever may be your ideas, I cannot see any reason for publishing the fact to the world.

Galton continued his statistical studies on prayers for years. When friends believed that ancient curses on church property affected inheritances by eldest sons, Galton gathered statistics showing that an equal percentage of eldest sons owned church and nonchurch property, and that the average length of tenure was identical. Divine intervention did not occur. His study was published in a 1909 issue of *Nature*.

Throughout his life, Galton worked on a number of strange inventions. Aside from the ventilating hat mentioned in the "Fact File," one of my favorites was a device with five interconnecting dials. The four-inch-long device actually employed five finger-operated keys. When a key was pressed, a pointer moved one division around one of five dials. Galton claimed, "It is possible by its means to take anthropological statistics of any kind among crowds of people without exciting observation, which it is otherwise exceedingly difficult to do."

Galton used another invention for compiling a "beauty map" of Britain that showed where to find the most beautiful women. The map was based on a statistical assessment of the number of beautiful, nondescript, and ugly women he saw on the streets of various towns. Galton wrote the following regarding his beauty map of the British Isles:

> Whenever I have occasion to classify the persons I meet into three classes, "good, medium, bad," I use a needle mounted as a pricker, wherewith to prick holes, unseen, in a piece of paper torn rudely into a cross with a long leg. I use its upper end for "good," the cross arm for "medium," the lower end for "bad." The prick holes are kept distinct and are easily read off at leisure. The object, place, and date are written on the paper. I used this plan for my beauty data, classifying the girls I passed in streets or elsewhere as attractive, indifferent, or repellent. Of

course this was a purely individual estimate, but it was consistent, judging from the conformity of different attempts in the same population. I found London to rank highest for beauty; Arberdeen lowest.

In addition to his beauty index, Galton worked on a boredom index based on the average rate of fidgets among people in an audience. In particular, while observing the heads of bored audiences, he noticed that heads waved back and forth with a frequency that quantified the degree of boredom. Attentive audiences sat upright and maintained a constant distance between heads.

> Many mental processes admit of being roughly measured—for instance, the degree to which people are bored, by counting the number of their fidgets. I not infrequently tried this method at the meetings of the Royal Geographical Society, for even there dull memoirs are occasionally read. . . . The use of a watch attracts attention, so I reckon time by the number of my breathings, of which there are 15 in a minute. They are not counted mentally, but are punctuated by pressing with 15 fingers successively. The counting is reversed for fidgets. These observations should be confined to persons of middle age. Children are rarely still, while elderly philosophers will sometimes remain rigid for minutes altogether.

Despite his intelligence, Galton lacked the natural ability to connect emotionally with people. For example, when he gave lectures, Galton could not be sure of the audience's reception by simply looking at their faces. Instead, Galton evolved a complex system of hand signals that his wife would use to tell him if he was speaking at the appropriate volume and with the proper speed. For example, Louisa would raise her left arm to one of two positions to indicate that he should raise or lower his voice. She would raise her right arm to signal speed adjustments. (Although many people have problems with public speaking, Galton's methods for dealing with these problems seem extreme.) During his talks, Galton had the curious habit of wearing a heart monitoring device beneath his suit, and I suspect this did little to increase the smoothness of his presentations!

Here are just a few additional examples of Galton's hyperpassion for measurement. He would place pressure sensors under his dining-room chairs in order to record body movements of his guests. He theorized that people in agreement, or who liked one another, would incline their chairs toward one another. He also would ask random people on the street how they liked the weather in order to assess the percentage of optimists and pessimists. When he had his portrait painted, he spent the time counting the number of brush strokes made by the artist. The results were reported in a 1905 issue of *Nature* titled "Number of Strokes of the Brush in a Picture."

In his 1894 paper "Arithmetic of Smell," published in *Psychological Reviews*, Galton described a device he designed that presented a controlled whiff of smell to either of his nostrils. He taught himself to associate two whiffs of peppermint with one of camphor, three whiffs of peppermint with one of carbolic acid and so on. He was then able to perform additions with the scents and use no visual or auditory associations.

Galton invented spectacles that gave divers clear vision under water. To perfect the spectacles, he experimented with various combinations of lens while in his bath until he found one that allowed him to read a newspaper under water! Galton wrote:

> I amused myself very frequently with this new hobby, and being most interested in the act of reading, constantly forgot that I was nearly suffocating myself, and was recalled to the fact not by any gasping desire for breath, but purely by a sense of illness, that alarmed me. It disappeared immediately after raising the head out of water and inhaling two or three good whiffs of air.

Galton observed people at horse races. As the horses neared the finish line, he noticed that the average hue of faces became a canvas of dark pink. A short note on this experience was published in an 1879 issue of *Nature*.

In 1906 he published "Cutting a Round Cake on Scientific Principles" in *Nature*. His aim was to develop a cutting pattern that would best preserve the freshness of the cake. A rubber band was used to keep the segments together.

Galton even developed complicated formulas for determining how much morning and evening tea to drink and how much water to use. His notebooks are filled to overflowing with arcane equations. Galton's experiment involved parameters such as the temperature of the water and teapot, the amount of tea, and the time allowed for brewing.

One of Galton's greatest passions was comparing the various races of humanity. For example, he suggested the relative worth of blacks and whites be measured by examining historical accounts of encounters between black chiefs and white travelers:

> The latter, no doubt, bring with them the knowledge current in civilized lands, but that is an advantage of less importance than we are apt to suppose. A native chief has as good an education in the art of ruling men as can be desired; he is continually exercised in personal government, and usually maintains his place by the ascendancy of his character shown every day over his subjects and rivals. A traveller in wild countries also fills, to a certain degree, the position of a commander, and has to confront native chiefs at every inhabited place. The result is familiar enough—the white traveller almost invariably holds his own in their presence. It is seldom that we hear of a white traveller meeting with a black chief whom he feels to be the better man.

Finally, I'd like to tell you about an intriguing feature of Galton's health involving "yo-yo sickness." Galton and his wife Louisa seemed to become ill like out-of-phase bouncing balls. For example, Louisa was sick when Galton was busily occupied away from home. When he returned and paid more attention to her, she would improve and he would sicken. Galton's symptoms appeared whenever he was prevented from working. Galton biographer D. W. Forrest notes:

> Every psychiatrist is familiar with the obsessive–compulsive patent whose constant temptation is to overwork and whose illness is exacerbated by a more leisurely routine. Like two yoyos out of phase, they would alternate into and out of sickness.

Modern forensic science owes much to Sir Francis Galton, and so do genetics, meteorology, and mathematics. The distinguished British scientist was brilliant beyond dispute, and he was as quirky as they come. His obsession with the notion that almost anything could be counted, correlated, and made into some sort of pattern makes him a good candidate for inclusion in this book.

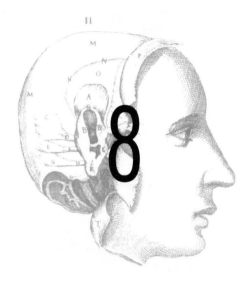

THE ICE MAN FROM CORNWALL GARDENS

PRODUCING A SINGLE BOOK ABOUT SUCH A MAN IS
NO EASY TASK, FOR IN HIS LIFETIME HE DID TOO
MUCH, THOUGHT TOO MUCH, AND WROTE TOO MUCH
THAT MATTERED. ANY SINGLE ONE OF HIS MAJOR
PROJECTS COULD BE EXPANDED TO FILL A LARGE
BOOK. HIS APHORISMS, HIS VITAL IDEAS ABOUT
IDEAS, WOULD FILL STILL MORE BOOKS.

DAVID LAMPE
PYKE: THE UNKNOWN GENIUS, 1959

THE CENTRAL FACT OF OUR EPOCH IS THAT
KNOWLEDGE HAS GROWN; MAN'S BRAIN HAS NOT.

GEOFFREY PYKE, 1931

📁 FACT FILE

- **Name:** Geoffrey Nathaniel Pyke
- **Born:** 1894, Cornwall Gardens
- **Died:** 1948, Hampstead (by suicide)
- **Nickname (given by Cavendish Laboratory scientists):** "The Ozzard of Whizz"
- **Occupations:** Inventor, philosopher, military tactician, foreign correspondent, advertising agent, financier, journalist, charity organizer, statistician, economist, broadcaster
- **Accomplishments:** Pyke had ideas for numerous inventions, controlled one-third of the world's supply of tin, and wrote a spell-binding book about his escape from a German prison. His most brilliant idea was the creation of aircraft carriers made of "pykrete," an incredibly strong form of ice. He also contributed to progressive theories of education for schoolchildren.
- **Some publications:** *To Ruhleben—and Back; a Great Adventure in Three Phases* (1917), "Politics and Witchcraft" (1934), "Latrines-for-Colonels Only" (memo, 1942), "Note on the Military Significance of Power-Driven Rivers" (1943), and "Organization of Thought" (1944)
- **Appearance:** Pyke was a tall man with a goatee that made his long head seem longer. His steel-framed spectacles rode absurdly high on his aquiline nose. His clothes were ludicrously crumpled. He also wore brightly colored spats because, as he said, "they can be worn for weeks *and* they obviate socks which I'd have to change much more often." Despite this, he was considered handsome and charming.
- **Religion:** Jewish but nonpracticing
- **Strange brain:** Pyke conducted most of his important military meetings while in bed because he was too busy to leave bed. He also exhibited hypergraphia (obsessive writing), and he filled notebooks with incessant ideas, scribblings, and newspaper cuttings. Pyke exhibited many unexplained periods of depression. His diet was often extremely homogeneous, sometimes consisting of only herring and biscuits. He detested socks. (I believe Pyke had undiagnosed temporal lobe epilepsy, as justified in the end of this chapter and in Chapter 16.)
- **Little-known distinction:** Pyke created a school that had Britain's first jungle gym.
- **Favorite quotes about Pyke:**

> "One of the most original, if unrecognized, figures of the present century." (*The Times*)

"His brain was a firework of ideas, some brilliant, some fantastic, but all new and highly unconventional." (Dr. Max Perutz, Cambridge University scientist)

"I regard him as one of the most original personalities of our time [with a mind like Einstein's]" (Lancelot Law Whyte, philosopher and scientist)

"Though little appreciated in his lifetime, Pyke will be counted among the great creative minds of the age." (Professor John Cohen, Manchester University psychologist)

"Everybody's conscience." (*Time*)

"He was the sort of man who would have invented the wheel." (Cyril Ray, *Manchester Guardian*)

"Pyke's wartime ideas are so dazzling they do not at the moment need to be discussed." (Winston Churchill)

• Favorite quotes by Pyke:

"Wisdom is like gold; it is useless if no one will accept it from you." (1931)

"I wonder if any of you think it is possible that one of the reasons why, apart from health, so little progress and investigation has been made in two and a half thousand years into the upbringing of the young is just that it would be uncomfortable to produce children greatly superior to ourselves. Do you think it possible that the reason why we have done so little, even to the extent of instituting research about it, is quite simply—that we don't want to?" (1931)

"For many years I had made a habit of turning jests upside down to look for the truth in them. For it is the concealed truth that makes the jest." (1931)

☞ THE STRAIGHT DOPE

Like many of the other strange brains in this book who lost one of their parents at an early age, Geoffrey Pyke's father died suddenly at age 44 when Pyke was five years old. Immediately after the death, Pyke's mother Mary gave him the unusual burden of functioning as head of the household. It was to be Pyke's job to figure out how he and his brothers and sisters would live in virtual poverty. Over the years, Pyke's mother complained incessantly about the lack of money but never did anything to help.

Mary Pyke sent her son to Wellington, a public school attended mostly by the sons of army officers. Mary should have been able to predict that her son would be mercilessly bullied at Wellington, not only because he was a Jew, but also because she made him observe a special diet and wear strange clothes. The Wellington boys never had a practicing Jew in their school, and they soon began to initiate pogroms, or "Jew Hunts," torturing Pyke with schoolboy pranks of one form or another.

Holidays back at home were like slipping from the frying pan into the fire. Pyke constantly had to assume the absurd role as head of the family and to protect his siblings from the unpredictable rages of his mother. His only respites were the times he visited his uncle's chemistry lab, where he would spend hours happily experimenting. Like Cavendish, discussed earlier, the young Pyke was curious about everything around him, and he was particularly fascinated by the world of technology and science.

After two years in Wellington hell, Pyke was removed and given a private tutor. Next he went to Cambridge to study law, and he returned home only when essential. On one rare occasion when his University friends visited him at home, Pyke was dressed in his mother's dress, amusing his brothers and sisters. After leaving home for good, Pyke never again practiced Judaism nor did anything that would remind him of his pitiful early life.

Not much is written of Pyke's siblings. Dorothy, Pyke's timid, gangling sister, escaped their unhappy home when she married a Christian and broke off all contact with her family. (She died a widow in a Surrey cottage, in 1957.)

When war broke out in 1914, Pyke suggested to *The Daily Chronicle* that he secretly go to Germany and send back reports. The newspaper agreed to pay his expenses but told him they could not be responsible for his safety. Pyke was soon sitting in German cafes listening to conversations and was eventually arrested as a spy. He was taken to Ruhleben, a prison camp, and placed in solitary confinement for 112 days. His cell had no electric light and was dark by midafternoon. The jailers wouldn't give him any books or paper, so he passed his time doing arithmetic in his head.

Day after day Pyke formulated plans for prison escape. He began to compile statistics on previously failed escape attempts, and noticed all these escapes were attempted at night. Pyke therefore decided to leave during the day when the afternoon sun shone directly into the eyes of a guard who inspected a store-hut on the camp grounds. He hypothesized correctly that if he were to crouch in the hut at exactly the right time, he would not be seen. The plan worked and he made the long walk to the Dutch frontier.

The Daily Chronicle was excited about Pyke's escape from Germany and published his memoirs. Pyke became an instant hero, and he even wrote a book about his escape. He began to give lectures on his experience and, while speaking at the Wellington school, he told his audience of schoolboys, "When things were at their worst in Germany, even when I was quite certain I'd be taken out and shot as a spy, I was never quite so unhappy, never so completely miserable as I'd been when I was a boy here at Wellington."

In 1918, Pyke married Margaret Chubb, a peaceful, intelligent, and pretty daughter of a physician. Pyke soon became a financier and explained how he chose his new profession with the following words: "I went into the City and spent a day watching men entering and leaving the Stock Exchange. All of them appeared ineffably stupid, and many of them were my relatives." Pyke believed he could do better, so he rented an office and started dealing in the commodities market. He ignored conventional methods and developed his own system after reading whatever literature he could find on the commodities market. Pyke was quite a frenetic businessman, directing every move and transaction as if it were a part of some Shakespearean play. Although not really necessary, he would post letters from cities he had not visited to distract others from discovering the real secrets of his speculations. Pyke's investments succeeded, and soon he was rich. At one point he controlled one-third of the world's supply of tin.

When Pyke and Margaret had a son David in 1921, Pyke decided that David should have the perfect upbringing. David should not be exposed to inhibiting or guilt-producing influences. Pyke studied all the available literature on education and formulated his own theories. So strong were Pyke's feelings that he created a school in a large house in Cambridge where he could educate David. Pyke placed ads in magazines to acquire quality teachers. "Preference will be given to those who do not hold any form of religious belief," Pyke wrote in the ad, "but this is not by itself considered to be a substitute for other qualifications." Pyke provided the school with all sorts of novel equipment. The children's see-saws had hooks on their undersides where children could place weights. In this way, the children would teach themselves about fulcrums without even being aware they were learning. Double-handed saws in the children's carpentry shop taught the children cooperation.

In Pyke's school, students became co-investigators with teachers. Teachers never forced instruction on a child. One small boy spent his time climbing beneath the floor and tracing the building's electrical and plumbing work. When a pet animal died, the children would dissect it to see how it worked and determine what had gone wrong. Students were never yelled at.

In 1928, Pyke lost all his money. It turned out that his economic charts were not infallible, and his speculations in commodities had grown enormous. His successful school had to be closed and his son sent to an ordinary school. (In 1938, David would be accepted to a United States medical research unit.) Pyke went into a deep depression, trying to figure out what went wrong with his investing techniques. To recuperate, he left his wife and rented a rickety shack in Surrey where he lived alone. His beard grew longer and longer as he became an emotional and physical wreck. In 1930, a small, attractive young woman came to live in the shack as secretary, housekeeper, and cook. It is not clear how Pyke managed to keep her and live on virtually no money.

Despite Pyke's new poverty, he was always an altruist and inventor. For example, he formed a world union of Christian leaders to denounce Hitler's racial polices. He arranged for poor Spanish hospitals to get a supply of microscopes. He designed sidecars for motorcycles that could carry hot food to the battle front. Food was heated by exhausts from the engine. He suggested to the Spanish that they could use pedal power to move their trains. In 1938, he even explained how sphagnum moss could be used as bandages to help troops in the Spanish Civil War.

In 1939, Pyke tried to avert imminent war by taking an opinion poll in Germany to show that most Germans did not want to go to war or persecute the Jews. In fact, Pyke thought Hitler would refrain from war if presented with the results of the public opinion poll. Of course, Pyke could not openly take such a poll so he hired British students to pose as golfing tourists in Germany. They were to strike up conversations and ask their questions as unobtrusively as possible. The young interviewers agreed to Pyke's plan, although they found his personal appearance (rumpled pajamas top for shirt, a boot lace for a tie) distracting. By August 21st, Pyke had ten interviewers working in Germany. They compiled the following statistics after they returned with their golf clubs, barely escaping with their lives:

- 35 percent of the Germans favored persecution of the Jews.
- 16 percent thought territorial conquest was worth war.
- 40 percent wanted war in order to regain those colonies which the Versailles Treaty had taken away from them.

Overall, the poll indicated that many Germans were against the war, but by the time Pyke's survey began to circulate, the war had already started.

From what I can tell by poring through what little literature exists on Pyke, he rarely saw his wife Margaret anymore, but his inventive mind kept generating ideas. For example, he suggested that microphones be placed on British balloons to alert aircraft crews to incoming enemy planes over the horizon. Admiral Mountbatten, the chief of British Combined Operations, hired Pyke as a scientific advisor. At his first meeting with Mountbatten, Pyke looked like a derelict. His collar was grubby, his pants looked like they had been trampled on by a herd of wildebeests. Mountbatten was too polite to say anything. In contrast to Pyke's appearance, Pyke had a decisive, grand demeanor. He relaxed in his chair and quickly announced to Mountbatten, "Lord Mountbatten, you need me on your staff because I'm a man who thinks. But my services will not be purchased cheaply."

Although his decrepit clothing revolted many of the military men, Mountbatten protected Pyke from them because Mountbatten respected Pyke's ideas. After World War II, Mountbatten would call Pyke "the most unusual and provocative man I have ever met." Pyke did suggest many ideas, many of which were interesting but impractical. For example, he suggested that commandos with dog-teams be sent to destroy the Romanian oil fields. The howling dogs would sound like wolves to the guards, who would be scared away. Pyke also suggested tying bottles of brandy to St. Bernard dogs so that the guards would take the bottles and get drunk. In addition, if bombers could start fires in the oil wells, British commandos dressed as Romanian firemen in fire engines could get in and stoke the fires by spraying them with incendiary bombs.

Pyke invented special attack-sleds to use in the snowfields of Norway. Sometimes the machines could move in a single file leaving behind one trail so that the enemy would never know how many sleds were ahead.

Some sleds bore torpedoes. Pyke also suggested that an insect egg-laying mechanism be added to the sled. "It would be possible," Pyke said, "to use a species of oviduct to lay small bombs under the snow while our machines are moving which would wreck the endless tracks of our pursuers."

To keep the enemy away from the sleds, Pyke suggested labeling the sled installations as if they contained secret Gestapo death-ray devices. In particular, he suggested they have the following posted on the door:

GESTAPO-RESEARCH-AND-DEVELOPMENT INSTITUTE
SPECIAL-DEATH-RAY DEPARTMENT
Entrance to this installation of the above-named Department
is strictly forbidden to all troops of the Wehrmacht whatsoever,
without a special pass countersigned by
the General Commanding the District,
under penalty of instant court-martial.
For your own safety, keep a minimum distance
of two kilometers from the installation.

Pyke also suggested that the sleds could be hid in stalls marked in German: "Officer's Latrine. For Colonels Only." This was supposed to keep hyper-obedient German soldiers from investigating.

Pyke eventually decided he needed a grander home from which to hold military meetings, so he climbed a ladder into a friend's home. (When his friend returned a day later, he found Pyke sleeping in the master bedroom.) When Pyke held his military meetings, he was too busy to put on clothes. Papers were scattered everywhere, and cigarette butts, maps, and empty beer and milk bottles littered the carpet.

When Pyke paced around the room, he continually tripped over footstools and small tables. He suggested to the owner of the house, "The floor should be a place for walking. Rooms would be a lot more sensible if furniture where suspended from the ceilings by ropes and only lowered when it was needed." The owner replied, "Geoffrey, I don't think it would be very attractive."

Pyke's attack-sled scheme was approved, given code-name "Operation Plough," and trials were started in America on various designs. Pyke traveled to Washington on a military mission but the U.S. military was disgusted by his disheveled appearance and ignored most of his advice until it was too late for the sleds to be used in war.

Pyke's most impressive project was code-named Habakkuk, and it involved the creation of huge aircraft carriers made of "pykrete," a substance he invented. (Habakkuk was an ancient Hebrew prophet and the name of the Old Testament book bearing his name.) Perhaps the name Habakkuk was chosen because the biblical Book of Habbakuk contains the phrase, "a work which you will not believe though it be told to you."

Pyke's wonderful pykrete was a wood pulp and ice mixture that had very unusual properties. For one thing, it was extremely hard to break. Also, it was very slow to melt. If a ship were to be made of pykrete, it would be unsinkable. Torpedoes could hit and do little damage. Any minor damage could be quickly repaired. Pyke also indicated how cooling pipes could circulate cold air through the ship to make sure it stayed solid. Cargo could be stored in holes carved in the sides of the ice ships. Pyke also developed methods for steering the ship.

Pyke claimed that by encasing the pykrete ships in wood, they could be made several times the length of the Queen Mary and help win the war with Germany. The pykrete ships could carry aircrafts and attack enemy ships by spraying them with super-cooled water, freezing them instantly on contact, and forcing them to surrender. (Super-cooled water is liquid below the freezing point of water.) Pyke also suggested that huge blocks of pykrete be used to create a wall around harbors while soldiers proceeded inland, spraying railways tunnels with super-cooled water to seal them. Courtrooms made of pykrete, and quickly assembled in Hamburg, could be the site for war crime trials. Perhaps at the end of the war, Pyke dreamed of sealing Hitler alive in his bunker.

One of Pyke's memos suggested that huge neon signs be affixed to dummy pykrete ships with the words, "Bomb me—I'm a dummy!" To save money, the decoy ships could be made of hacked off pieces of icebergs rather than real pykrete.

The details of Pyke's invention of pykrete are fascinating. Pure ice has a molecular structure similar to pure concrete, but ice has a lower tensile strength than concrete. When small particles like the chips used in paper manufacture are added, the ice has remarkable strength. The crush resistance of ordinary ice is between 250 and 1,300 pounds per square inch. Pykrete has a crush resistance of greater than 3000 pounds per square inch. A one-inch column of pykrete can support an automobile! The wood pulp also makes the pykrete extremely stable at high temperatures. It turns out that a four percent (by weight) solution of wood pulp in water produces a suspension the consistency of porridge, and when frozen is quite strong. A 14 percent sawdust suspension is like wet concrete in consistency but much stronger than concrete.

Hit pykrete with a hammer, and it does not crush! Fire a .303 bullet at it, and it penetrates only 6.5 inches. Pykrete is so ductile that it can be turned on a conventional wood-working lathe. Pyke's colleagues bombed pykrete. Torpedoed it. Attacked it with incendiary material. And yet, pykrete withstood the barrage like an impenetrable shield.

Lord Mountbatten was so impressed with the pykrete that he ran to where Winston Churchill was staying and dropped a block of pykrete into Churchill's hot bath water. The pykrete did not melt.

Next, Churchill showed pykrete to President Roosevelt placing pykrete cubes into scalding water, where they did not melt. Mountbatten also showed pykrete to allied chiefs of staff in Canada. To demonstrate the awesome nature of pykrete, Mountbatten suddenly pulled out his gun and shot at it. The bullet merely bounced off the pykrete, nearly killing an admiral.

Just how practical were all of Pyke's suggestions? Although the use of super-cooled water was impractical for certain technical reasons, there was no reason that pykrete ships could not be built. In fact, the U.S. and Canada were so impressed with the idea of assembling pykrete warships that Operation Habakkuk went ahead. A 60-foot-long, 1000-ton pykrete ship was built in one month on a Canadian lake and never melted through the hot summer.

Pyke and his military experts soon had a plan for producing battle ship Habakkuk (2000 feet long and 2,200,000 tons—26 times that of the Queen Elizabeth). The Habakkuk, however, never had a chance to be tested in war because the Normandy landings made the use of ice ships unnecessary. Had the huge ship been built, some scholars believe she would have been "the second most spectacular device of the war, out-shadowed only by the atomic bomb."

Pyke was later granted the right to patent pykrete in the U.S., but he never bothered to file a patent. Despite the fact that the pykrete ships were never used in the war, Pyke continued to flood Lord Mountbatten with new inventions. One of my favorites was a pipeline for pumping equipment and soldiers from ships to shore or across difficult country. He called these water pipes "Power-Driven Rivers", which reminds me of Tesla's idea of using fluid-filled pipes to carry mail from the U.S. to Europe, discussed in Chapter 1. "The idea of transporting human beings inside the pipe," Pyke admitted, "has a slightly imaginative and speculative quality about it." Pyke worked out many technical problems, such as how to supply oxygen, but he knew that claustrophobia might pose a psychological problem, and men would therefore be more comfortable traveling in pairs. Also for psychological reasons, the containers should have lights, and the passengers should take small doses of barbiturates. Pyke commented:

> For long journeys, provision would have to be made for micturition and defecation. The former of course presents no special difficulty. But the latter presents a problem which it would be desirable to solve perhaps for psychological rather than for physiological reasons. However, should it prove insoluble, it would have to be accepted, perhaps not for the first time, as one of the drawbacks of war. The whole experience of riding in a pipe, however, should be far less unpleasant, and take very much less time to become used to, than parachute jumping, or getting bombed.

Pyke's inventive ideas continued to gush like a fountain. When the war was over, Pyke suggested that Europeans use pedal-powered machines, because Europe was manpower rich but energy poor. He worked out how complex assemblies of pedalers could move tractors, trucks, and trains. He even hired a boy to pile bricks all day in order to determine that a human can move in a day 20 times as much as he can carry. He believed that a world organization should be formed to help countries integrate muscle-power equipment into their economies.

Pyke's final days were spent in a room in Hampstead, where he often worked in bed so as not to waste time getting up and dressing. At Hampstead he developed theories on how to improve the world economy, and he formulated all kinds of guiding principles for future civilizations. So addicted was he to his bed that he installed an electric kettle that he could switch on and off without leaving his bed. He elevated the foot of the bed because he thought it would help his circulation. Medicine bottles, cigarette butts, books, file folders, and scraps of herring wrappers littered the floor. (He loved herring.)

Pyke was in principle a Marxist, but told a young friend that the friend was wasting too much time on politics when he should be concentrating his effort on his work. "If you need a second occupation," Pyke told him, "well, I suggest women. Then use whatever energy you have left for doing what you consider your civil duty."

Pyke became concerned with the problem of health and industrial production, and he began to question over what period of time nations should make plans. In fact, Pyke filled notebooks with algebraic equations in order to determine the length of time over which nations should make monetary investments in public health.

The winter of 1948 was a particularly cold and dreary one. There were coal and electricity shortages. Pyke continued to work on equations and seek psychological and philosophical truths to improve the economic state of humankind. But for no apparent reason, on Saturday evening, February 21st, Pyke shaved off his beard. Next he grabbed a bottle of sleeping pills.

His hypergraphia (obsessive writing) stayed with him to the end. He furiously wrote page after page, but they offered no reason for his suicide. They were private letters to his wife, son, and a psychologist advising him on nursing recruitment. Pyke took the pills and continued to write until his handwriting became illegible.

No one knew why Pyke committed suicide. At Pyke's request, he was cremated without ceremony. No gravestone was ever erected.

The Times in its obituary on Pyke wrote, "The death of Geoffrey Pyke removes one of the most original if unrecognized figures of the present century. Unfortunately, Pyke, although he spent himself on endless memoranda, published little or nothing, and his ideas have affected only those few who were associated with his schemes. On these, however, he made a deep and lasting impression, and it is to be hoped that his influence as a pioneer and an organizer of thought will continue and spread."

The *Manchester Guardian* wrote, "In spite of the publicity the Admiralty gave to Habakkuk, Geoffrey Pyke was not known to a wide public. His tremendous dignity, triumphing always over a sometimes almost ludicrous shabbiness, was all his own. With the death of Geoffrey Pyke, Britain has lost one of the greatest and certainly the most unrecognized geniuses of all time. Imagination was, indeed, his chief characteristic, and, indeed, part of his undoing."

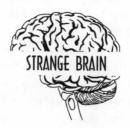

STRANGE BRAIN

The public knew little of Pyke because he was too busy working on his projects to spend any time promoting himself. He always believed that good ideas would prevail if they were approached with a "ferocity of purpose." His ferocity was obsessive. As mentioned in the previous section, he held meetings with Britain's busiest military planners in his bed because he was too busy too get dressed. Empty bottles littered his bedside because going down the hall wasted too much time.

His courtship to Margaret was equally unusual. When he was invited to dine with Margaret and her cousins, he showed up in a crumpled dinner jacket. The jacket was so wrinkled it looked as if it were tossed around by a pack of hyenas in heat. Rather than wearing reasonable shoes, he chose the dusty hiking boots that he had worn while escaping from Ruhleben prison. He slouched and pointed his long legs forward with his boots thrust insolently at his host.

Margaret asked when Pyke would introduce her and her family to his mother. "Never," Pyke answered. He did tell Margaret that their courtship was the first time he had ever been happy. "You've never really been happy before?" she asked. "Never that I can recall," he replied. "I've learned to live without needing happiness."

During his marriage, Pyke had unexplained periods of depression, during which he would shut himself away for hours or leave for days without explaining to Margaret where he had been. While he lived with Margaret, he acquired numerous books which he rapidly absorbed. One day Margaret found him poring over a manual for a Model T Ford, which he said was most interesting. When she asked him how he could find it interesting considering he didn't know how to drive, Pyke's response was, "What's that to do with it?" (This response isn't too strange considering his gift for mechanical design.)

While living at his shack with his secretary, Pyke began to keep weird hours, sometimes reading all night. At his bedside he kept as many as 60 books, from economic treatises to detective stories. Idea after idea flooded him on how to improve the world. Pyke filled notebooks with incessant scribblings and newspaper cuttings. Pyke biographer David Lampe writes:

> To transcribe his pages of scratchy, unruly scrawlings, Pyke dictated for long hours to his secretary who pecked patiently at the typewriter. When this seemed altogether too slow, Pyke would take over the machine so that she could dictate to him. New idea bred new idea, and each new plan flared up so quickly that he could not capture it on paper before the next

scheme seemed to reveal itself to him. Hardly anything seemed to get finished for always there was the new thought to chase down, the newer plan to develop because it seemed more important than the one that had preceded it.

Whenever Pyke took a break to visit friends, he would return to his shack and moan, "Tonight I talked too much! I threw away ideas I'll never be able to remember again! Everything must be put on paper right away! Don't let me talk so much! Make me discipline myself!"

Pyke would signal to his secretary that a depression was about to envelope him by humming, and she would learn to stay out of his way. Pyke also developed some strange eating habits. There were some nights that Pyke became so ravenously hungry that he devoured a week's supply of bread in one sitting. Recall that in Chapter 3 we saw another orgy of eating by Samuel Johnson.

Once while at a butcher's shop, Pyke spied a box of old, disgusting, gnawed meat. He asked the butcher what he could buy for a tuppence a pound. When the butcher said that he did not sell food that way, Pyke responded, "Come, come, my good man, you must have *something* for a tuppence a pound. What about that?" Pyke pointed to the box of scrap meat. The butcher responded that the meat came from inside the guts of the slicing machine after the machine was cleaned. "It aint fit to eat," the butcher said. Pyke replied in a shocked tone, "Do you mean that you sell food unfit for human consumption?" After some more bickering, Pyke bought the shredded meat for the price he offered. At other times, Pyke lived on a diet consisting almost entirely of herring and broken biscuits.

Pyke's continuous outpouring of ideas and writings reminds me of the extreme output of some artists with temporal lobe epilepsy (paroxysmal malfunction of brain cells in the lower middle of the brain)[1] and obsessive–compulsive disorder. The term for such an excessive compulsion to write or draw is called "hypergraphia." In Chapter 16, I discuss the link between temporal lobe epilepsy, creativity, and even religious visions in such artists and writers as van Gogh and Dostoevsky. Pyke himself was aware of a possible epileptic component to his creativity and commented in 1931:

I felt as epileptics must do when they are about to have an attack: a condition of generalized tenseness. I knew that a revelation was coming into my mind, and that details were falling fast into place. I concentrated on the horizon and in another breath or two my mind was quite clear and I felt as loose and content as I imagine people do when they have received what they imagine is a religious revelation.

My favorite example of Pyke's hypergraphia occurred when he called Headquarters at five in the morning to dictate to a junior officer a memo to Lord Mountbatten. Pyke began to dictate and the officer began to write. An hour later, Pyke was still dictating and the officer was furious!

Pyke alienated officers in other ways. One day he was walking with a brigadier who had just had a difficult day. "Mine is a terrible job at times," the officer lamented. "Some days I feel like chucking it in."

Pyke smiled back, patted the officer on the shoulder, and said, "Why don't you?"

When Pyke later recounted the event to a friend, Pyke wondered why the officer seemed offended. "What was I to say to him?" Pyke asked his friend. "The suggestion had been his."

Toward the end of his life, low blood pressure kept Pyke in bed all the time. His diet was very poor because he was too busy to worry about food, and he insisted on living like a pauper. Sometimes he had a single chop for his evening meal. Yet Pyke continued to write and write and write. Here are just a few examples of his bedside wisdom:

There wont be any future wars. Not if they don't begin in 20 years anyway.

The thing to remember is never to do what you feel you ought to want to do. Do things because you ought—or perhaps you want to. But never because you ought to want to. Never assume that if people do something different or want something different, that their way is necessarily wrong. It's just different.

Never get married. You'll find a girl who'll be prepared to sleep with you without marrying you—though she won't say so! Almost any affair would do you good, provided it isn't wholly unemotional. The prostitute relation, I mean. But you must solve this problem because otherwise you'll find that your frustration will seriously interfere with your work. Of course, if you have a girl friend, that will take up a lot of your time in other ways.

After Pyke's death, Lancelot Law Whyte, a physicist and philosopher who helped pioneer jet aircraft, said, "Pyke's method was to approach any problem on the widest possible basis, taking as little as possible for granted, in order to find the best line of attack. With Einstein, there is his theory. With Whittle,[2] there are the jet planes howling across the sky. Pyke's genius was more intangible, perhaps because he produced not one, but an endless sequence of ideas, each of which in turn obsessed him passionately, and each of which was lit with strange fires of his mind."

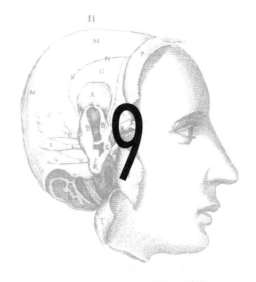

THE HERMIT FROM MONTANA

THE MAJORITY OF PEOPLE ENGAGE IN A
SIGNIFICANT AMOUNT OF NAUGHTY BEHAVIOR.

UNABOMBER MANIFESTO

FREEDOM MEANS HAVING POWER; NOT THE POWER
TO CONTROL OTHER PEOPLE BUT THE POWER TO
CONTROL THE CIRCUMSTANCE OF ONE'S OWN LIFE.

UNABOMBER MANIFESTO

LET F BE OUR FINITE SKEW FIELD, F^* ITS
MULTIPLICATIVE GROUP. LET S BE ANY SYLOW
SUBGROUP OF F^*, OF ORDER, SAY, p^α. CHOOSE AN
ELEMENT g OF ORDER p IN THE CENTER OF S. IF
SOME $h \in S$ GENERATES A SUBGROUP OF ORDER p
DIFFERENT FROM THAT GENERATED BY g, THEN g
AND h GENERATE A COMMUTATIVE FIELD
CONTAINING MORE THAN p ROOTS OF THE EQUATION
$x^p = 1$, AN IMPOSSIBILITY. THUS S CONTAINS ONLY

ONE SUBGROUP OF ORDER *P* AND HENCE IS EITHER
A CYCLIC OR GENERAL QUATERNION GROUP.

T. J. KACZYNSKI
"*ANOTHER PROOF OF WEDDERBURN'S THEOREM*"

 **FACT FILE**

- **Name:** Theodore Kaczynski (a.k.a. "Unabomber")
- **Born:** May 22, 1942, Chicago, Illinois
- **Occupation:** Mathematician
- **Achievement:** Kaczynski wrote several brilliant papers on the mathematical properties of functions in circles and boundary functions. (His IQ was approximately 170.)[1]
- **Marital status:** Never married; celibate
- **Strange brain:** Excessive (pathological) shyness, "fascination with body sounds," metronomic habit of rocking, and frequent concerns about germs, infections, and other health matters. Possible trazodone-induced priapism.[2] His room at Harvard stunk of rotting food and was piled high with trash. After teaching two years and publishing mathematical papers that dazzled his peers and put him on a tenure track at one of the nation's most prestigious universities, he suddenly quit, spent nearly half his life in the woods, and is suspected of killing three strangers and injuring 22 others. Throughout his life, Kaczynski found it painful to make errors and compulsively corrected minor errors in others. He shut himself up in his bedroom for days at a time and seemed incapable of sympathy, human insight, and simple connections with people.
- **Standard dress:** At Harvard he wore "a kind of unpleasant plaid sports jacket and a tie that didn't go with it" (Classmate Richard Adams, now an investor).
- **Residence as an adult:** A 10-by-12-foot shack in the woods of Lincoln, Montana
- **Religion:** No religion. Clearly not interested in Christianity.
- **Despised:** His parents
- **Favorite quotes about Kaczynski:**

  > "Most classmates regarded him as alien, or not at all."
  > (Robert McFadden, *New York Times*, 1996)

"He was a person who seemed to be capable of closing doors on things, on people, on stages of his life. Cutting himself off was part of what he was about." (David Kaczynski, brother)

"He'd have to take a bath if he worked here." (Jay Potter, owner of a store to which Kaczynski applied for a job)

"He was never seen as a person, an individual personality. He was always regarded as a walking brain, so to speak." (Jerry Peligrano, classmate)

"I don't recall 10 words being spoken by him. I do remember trombone blasts and foul odors from his junkyard room." (Patrick McIntosh, Harvard)

"A lot of mathematicians are a little bit strange in one way or another. It goes with creativity." (Professor Peter Duren, Kaczynski's graduate school teacher)

- **Favorite Book and Author:** Kaczynski's favorite book and author was *The Ancient Engineers* by L. Sprague de Camp, according to the librarian in Lincoln, Montana. (*The Ancient Engineers* describes the early but major technological achievements that took place during the evolution of civilization, such as the construction of the Egyptian pyramids, the Parthenon, and the Great Wall of China.)

☞ THE STRAIGHT DOPE

THE CASE OF THE UNABOMBER PROBABLY BELONGS MORE TO THE HISTORY OF SOCIOPATHIC PSYCHIATRY THAN TO THE HISTORY OF POLITICAL IDEAS.

NANCY GIBBS ET AL.

THUS EVERY SYLOW SUBGROUP OF $F*$ IS CYCLIC AND, $F*$ IS SOLVABLE. LET A/Z (WITH $Z \subset A$) BE A MINIMAL NORMAL SUBGROUP OF $F*/Z$. A/Z IS AN ELEMENTARY ABELIAN GROUP OF ORDER P^K (P PRIME) SO SINCE THE SYLOW SUBGROUPS OF $F*/Z$ ARE CYCLIC, A/Z IS CYCLIC. . . . THUS WE HAVE

PROVEN THAT *A* IS CONTAINED IN THE CENTER OF
F*, A CONTRADICTION.

T. J. KACZYNSKI
"ANOTHER PROOF OF WEDDERBURN'S THEOREM"

Ted (Theodore) Kaczynski was born in Chicago in 1942. His father, the son of Polish immigrants, was a sausagemaker. By all accounts Kaczynski's father was a gregarious outdoorsman who valued education and was hardworking, thoughtful, and kind. Kaczynski's doting mother, also a child of Polish immigrants, read issues of *Scientific American* to him when he was only 10 years old, and she appeared to be quite proud and fond of him as a boy.

Like many others in this book, Kaczynski stood out as being since childhood particularly intelligent. He quickly went through high school, skipping his junior year, but made only passing gestures at social contact. During this time he enjoyed creating battery-powered devices that exploded using potassium nitrate. He constructed these small explosives using materials he could find in the house or local hardware store.

While in high school, Kaczynski also worked on devices that employed paper and ammonia. For example, one device required him to carefully twist a small piece of paper in the middle and cup the two ends. In one end, he placed a few drops of ammonia. In the other end, he placed iodine. He closed the two ends, gave it to a high school classmate and told her to untwist the middle.

When the chemicals ran together, the small device exploded in her hands with a harmless pop. Kaczynski laughed. So did the surprised girl. When she told Kaczynski that he was going to get suspended, he replied, "No, I'm not. I'm too smart."

When he was only 16, Kaczynski went to Harvard University where he shared his living quarters with several fellow students. He rarely interacted with or even spoke to them. According to one of his suite mates, Kaczynski's room was piled with trash two feet deep and underneath it all were what smelled like unused cartons of milk. The stench was horrible. On the occasions Kaczynski ate in the dining hall, he almost always ate alone, usually at one of the corner tables.

Kaczynski graduated from Harvard by the time he was 20, but had no friends. He next went to the University of Michigan, where he got his master's (1964) and Ph.D. (1967). Despite almost five years' residency at the University of Michigan, he left no pictures nor yearbook entries. Nevertheless, he left his mark in a different way. His mathematical papers were distinguished, obviously the work of a genius, and perhaps understandable by only a dozen people in the country at the time!

Next he went to the University of California at Berkeley to teach. (I'll go into more detail in the "Strange Brain" section.) Here Kaczynski was an assistant professor on a tenure track at the world's premier math department, an enviable position for any young mathematician. But after two years he suddenly resigned without giving a reason. The chairman of the mathematics department at the university described him as "almost pathologically shy."

In 1971 Kaczynski traveled to Montana, buying land, building a tiny house, and living on what he could grow or kill. Kaczynski's cabin in the woods had no phone, plumbing, or electricity, but it did have two walls filled floor to ceiling with Shakespeare, Thackeray, and purported bomb manuals. Sometimes he would stay inside his cabin for weeks at a stretch. At other times he would ride his bike, dressed in black, into Lincoln to use the phone, or read endlessly at the library (*Scientific American, Omni*). Sometimes he asked for special books that the small library had to order from afar. For example, he read German classical literature and wanted it in German.

Kaczynski would stuff both books and groceries into his dingy backpack and pedal back to the cabin, reading by the light of his own homemade candles. Occasionally, he would buy Spam, canned tuna, and flour from the local grocery store and also load them into his backpack. He did odd jobs now and then but apparently subsisted on a few hundred dollars a year, with plenty of free time for what many[3] believed to be his growing vocation: the disruption of the industrial society he had left behind. His 18-year crime spree entailed the bombing of numerous strangers living in the United States.

Kaczynski is the "Unabomber," an individual who mailed various killing devices patched together from lamp cords, bits of pipe, recycled screws, and match heads. He fits well with our assortment of strange geniuses: a mathematical wizard who rejected society and social discourse for the life of a hermit.

Let us study separately what we know of the Unabomber. The Unabomber's first bomb was sent to a professor at Northwestern University in 1978. A year later, a second bomb was left at the university, injuring a graduate student who opened it. After that they were sent to an airline executive, the computer science departments at Vanderbilt and Berkeley, and a University of Michigan professor. The FBI dubbed him the "Unabomber" because of the bomber's favorite targets: universities and airlines.

The Unabomber was particularly meticulous in his designs. For example, in order to avoid being caught, he made his own explosives out of commonly available chemicals. He scraped the labels off batteries so they could not be traced, used stamps long past their issue dates, and wires that were out of production. He left no fingerprints. Despite all of this, he also used distinctive handmade components when store-bought parts would have worked better and been harder to trace. There were always clues and inside jokes—his obsessive trademark usually having something to do with wood that he polished and varnished. We do not yet know if there is a connection between his use of wood and a leaf symbol that surrounded the bomb sent to one of his targets: Percy Wood, president of United Airlines. The fact that the Unabomber never injured himself with all his bombs, and the fact that the pieces were numbered, is also evidence of his patience and careful, meticulous nature.

Altogether the Unabomber killed three strangers and injured 22 others. Toward the end of his killing spree, the Unabomber was writing to the *San Francisco Chronicle,* threatening to blow up an airplane out of Los Angeles. He promised to stop the killing if the *New York Times* and the *Washington Post* would publish his magnum opus, his Manifesto, a 35,000-word diatribe against industrial society and modern civilization.

On April 3, 1996, Theodore John Kaczynski was taken into custody and subsequently charged with some of the Unabomber crimes. When FBI agents arrested Kaczynski in his little shack in the Montana woods, they purportedly found numerous books and ten three-ring notebooks full of data, diagrams, and test results—the careful professor's lab report on the quest for the perfect bomb.

Ted Kaczynski was docile and impassive at his arraignment in Helena, Montana. Like some other serial killers, Kaczynski had been a quiet boy, never got into any major trouble as a boy, and, as a possible criminal, left no prints. He made few friends right up until the moment he vanished into the woods and was arrested.

The details of Kaczynski's 1997 trial brief looked like, this:

In The United States District Court
for The Eastern District of California

United States of America,
Plaintiff

v.

Theodore John Kaczynski
aka "FC"
defendant.

CR No. S-CR-S-96-259 GEB

Violations: 18 U.S.C. Section 844 (d)—Transportation of an Explosive With Intent to Kill or Injure (4 counts): 18 U.S.C. Section 1716—Mailing an Explosive Device With Intent to Kill or Injure (3 Counts): 18 U.S.C. Section 924 (c) (1)—Use of a Destructive Device in Relation to a Crime of Violence (3 Counts)

In January of 1998, Ted Kaczynski plead guilty to charges he engaged in a lethal 17-year campaign of terror in exchange for a sentence of life in prison without possibility of parole. The Kaczynski trial had three false starts, and presiding Judge Garland Burrell clearly blamed Kaczynski for maliciously manipulating the criminal-justice process in order to delay the court proceedings.

After Kaczynski apparently attempted to strangle himself with his underwear while in his jail cell, he asked the judge to dismiss his lawyers; Kaczynski wanted to provide his own defense. At Burrell's order, Kaczynski then underwent five days of mental competency test by federal psychiatrist Dr. Sally Johnson to determine if he could stand trial and defend himself. While Johnson found Kaczynski competent to stand trial, she confirmed defense lawyers' contention that he suffered from paranoid schizophrenia, and delusions that he was harassed by society. Kaczynski had been battling for months with his lawyers over their plans to protray him as a high-functioning paranoid schizophrenic who should not be executed if convicted.

On a personal note, during the height of concern about the Unabomber, my employer began intercepting and X-raying suspicious boxes that were mailed to me—I was thought to be a candidate for the Unabomber due to my reputation as an author writing on technology issues. All of this made me wonder how a mathematical genius such as Ted

Kaczynski could also be mad, obsessive, delusional. . . . It was at the height of Unabomber terror and fear that I started writing *Strange Brains and Genius*.

STRANGE BRAIN

"TED HAD A SPECIAL TALENT FOR AVOIDING RELATIONSHIPS BY MOVING QUICKLY PAST GROUPS OF PEOPLE AND SLAMMING THE DOOR BEHIND HIM," SAYS PATRICK MCINTOSH, ANOTHER OF THE SUITE MATES. KACZYNSKI'S ROOM WAS A SWAMP; THE OTHERS FINALLY CALLED IN THE HOUSEMASTER, THE LEGENDARY MASTER OF ELIOT HOUSE JOHN FINLEY, WHO WAS AGHAST.

TIME, APRIL 15, 147(16): 38–46.

THE CABIN

ONE OF MY SUITE MATES, I RECALL, SEEMED MORE INTERESTED IN INSECTS THAN PEOPLE.

PROFESSOR MICHAEL ROHR

Ted Kaczynski was a mathematical genius who rose swiftly to academic heights even as he became an emotional cripple and loner like Henry Cavendish, the great scientist discussed in Chapter 6. (Recall that Cavendish was so shy that he ordered his female servants to remain out of sight or be fired.) Kaczynski's 25-year, self-imposed exile in the Montana woods was particularly appropriate for this man who had always been alone. The May 26, 1996, *New York Times* noted that the cabin "suited this genius with gifts for solitude, perseverance, secrecy and meticulousness, for penetrating the mysteries of mathematics and the dangers of technology, but never love, never friendship." The remoteness of the cabin was probably as much a means of limiting access of others to him as it was a symbol of freedom.

Lincoln, Montana, seemed to be a place where people had a high tolerance for eccentricity, allowing Kaczynski to flourish in his own way just like the Victorian times allowed other strange brains in this book to flourish. While in the woods, his inability to cope with people decreased, while his obsessions increased to the point that even his family did not recognize him on the rare occasions they saw him or in the hundreds of letters they received until he finally stopped writing them. He had come to believe his parents had cared more about his brain than his happiness, and when David, his seven-year younger brother, married, Kaczynski considered it an act of betrayal. For example, years later David remarked,

> When I told him of my decision to marry, he decided to end his relationship with me, end communicating with me. He had gotten the notion that my fiancee was a manipulating female who was using me. It may have just been terrible for him to think I would rejoin society.

In many ways Kaczynski's stay in the woods was a return to activities from childhood: camping in the forests with his family, setting off numerous small bombs while in high school. Even his wood carving interests were part of his early childhood.

CHILDHOOD

> IF YOU HAD ANY BRAINS, YOU WOULD HAVE
> REALIZED THAT THERE ARE A LOT OF PEOPLE OUT
> THERE WHO RESENT BITTERLY THE WAY TECHNO-
> NERDS LIKE YOU ARE CHANGING THE WORLD, AND
> YOU WOULDN'T HAVE BEEN DUMB ENOUGH TO OPEN
> AN UNEXPECTED PACKAGE FROM AN
> UNKNOWN SOURCE.
> LETTER FROM THE UNABOMBER TO HIS VICTIM,
> YALE PROFESSOR DAVID GELERNTER

How did it all begin? We can speculate on causes of Kaczynski's aberrations, although I don't think there is any one environmental factor or sequence of events that led to his strange behavior. However, one distressing event in his early life comes to mind. At the age of six months he was hospitalized for several weeks after suffering an allergic reaction to a drug.[4] During that time, the hospital did not allow his parents to hold or hug him, because he was kept in isolation. When he came home, they found him listless and withdrawn. His brother recounts the family story:

> After Teddy came home, he became very unresponsive. He had been a smiling happy jovial kind of baby beforehand, and when he returned from the hospital, he showed little emotions for months.

Could some peculiar event at the hospital, or the withdrawal, have triggered some of his problems later in life? In light of this early denial of human contact, it is strangely ironic that one of the Unabomber's early targets was James McConnell, a professor of psychology at the University of Michigan who eventually became well known for researching the benefits of sensory deprivation for autistic children.

Another isolated event occurred when Kaczynski was seven years old, at which time he was left alone to sob in a hospital lobby while his father and grandmother went to the maternity ward where his brother had been born. However, it's hard to believe these isolated incidences were of such magnitude that they would contribute to a Unabomber's bizarre and deadly behavior later in life.

Despite some possible traumas in childhood, there is no doubt that Kaczynski is a genius. When he was six years old, child psychologist Ralph Meister gave Kaczynski an IQ test that showed him to be in the range of 160 to 170. (Psychologists consider scores over 140 to be in the genius level.) Kaczynski was more than bright. He possessed rare intellectual gifts. Dr. Meister was equally amazed when Kaczynski suddenly asked him for a copy of *Robert's Rules of Order*, a dry handbook of parliamentary procedure.

The young Kaczynski soon became an obsessive reader. Once when he was 10 or 11 and going on a camping trip with his family, Kaczynski ran to a neighbor's house asking that she care for his pet bird. When she noticed a book under his arm, she thought it was a story book and asked him why he would bring a book on a camping trip. Kaczynski, in fact, was carrying *Romping Through Mathematics, From Addition to Calculus*. His response to the neighbor was, "I have to learn this."

Kaczynski's father loved the outdoors and took his boys in the woods for a week at a time. Sometimes they tried to live only on whatever they could gather or kill. They even learned from their father the proper way to eat a porcupine.

Kaczynski gradually began using his intelligence like a knife. For example, when a three-year-old girl from the neighborhood mispronounced the word "grasshopper" he stopped on the sidewalk and lectured the perplexed child on the scientific classification of grasshoppers.

Even his aunt recalls Kaczynski's rapier wit and absence of compassion or feeling for others. Once when she was visiting him and 11-year-old Kaczynski did not talk to her, Kaczynski's father asked him, "Why don't you talk to your aunt?

Kaczynski's answer was, "Why should I? She wouldn't understand me anyway."

Kaczynski's precocious mind was in evidence even at play. For example, he sometimes joined the neighborhood women playing Scrabble. One neighbor noted,

> Within minutes, he had all of us beaten. We were not dumb. Teddy was brighter. At 12, he had a better vocabulary than we did.

HIGH SCHOOL

WE WOULD NOT WANT ANYONE TO THINK THAT WE HAVE ANY DESIRE TO HURT PROFESSORS WHO STUDY ARCHAEOLOGY, HISTORY, LITERATURE OR HARMLESS STUFF LIKE THAT.

UNABOMBER'S LETTER TO THE NEW YORK TIMES

Kaczynski progressively avoided human contact and began to have discipline problems in school. During high school, his parents wondered why he was unhappy despite his intelligence. They felt it might be due to the fact he was so much brighter than his classmates. A high-school math teacher considered Kaczynski one of his brightest students but also remarked, "He drove his teachers up the wall." Perhaps the teacher was referring to devilish pranks, such as mixing compounds of ammonia and iodine that would pop loudly but harmlessly in a classroom and create violet smoke.

You have read about many of the strange geniuses in the book who had problems relating to women. As far as I can tell, Kaczynski had no substantive relationships with a woman during his life. This lack of romance was not because Kaczynski lacked interest. For example, he dated a girl a few times after high school graduation but broke off any possibility of a continued relationship with a refusal to endure her Catholic ideas.

While in high school, a school wrestler asked Kaczynski's advice on making a better bomb. Kaczynski obliged. When the bomb went off in chemistry class, it blew out windows and inflicted temporary hearing damage on a girl. Even though everyone was reprimanded, Kaczynski seemed unfazed. In fact, he continued to enjoy setting off small explosions that echoed across the neighborhood and sent garbage cans flying.

His painful shyness and his developing fascination with body sounds continually set him apart from his peers. Most of his classmates started to either regard him as an alien, as a walking brain, or as someone not worth paying attention to. One of his classmates, Patrick Morris, said:

> Ted was technically very bright, but emotionally deficient. While the math club would sit around talking about the big issues of the day, Ted would be waiting for someone to fart. He had a fascination with body sounds more akin to a five-year-old than a 15-year old.

His obsession with sounds would escalate later when he was at Harvard. There his annoying trombone blasts in the middle of the night bothered his suite mates. Paradoxically, his brother notes that Kaczynski never liked TV or rock and roll, and loud noises infuriated him. Interestingly, the Unabomber wrote in the manifesto,

> Technology exacerbates the effects of crowding because it puts increased disruptive powers in people's hands. For example, a variety of noise-making devices: power motors, radios, motorcycles etc. If the use of these devices is unrestricted, people who want peace and quiet are frustrated by the noise.

Kaczynski spent most of the time in high school in awkward solitude and was once stuffed into a locker by a bully. He did have a few hobbies, including wood carving and pulling pranks.

His parents and teachers decided to allow Kaczynski to skip his junior year of high school and sent him off to Harvard at age 16. Ted seemed to revel in his intelligence, but he may have felt valued solely for his brain, the way a pretty model may feel liked purely for her beauty.

The Unabomber in his manifesto makes a provocative comment on the pressure placed on young, intelligent children:

> The system needs scientists, mathematicians, and engineers. It can't function without them. So heavy pressure is put on children to excel in these fields. It isn't natural for an adolescent

human being to spend the bulk of his time sitting at a desk absorbed in study. A normal adolescent wants to spend his time in active contact with the real world.

Among primitive peoples the things that children are trained to do tend to be in reasonable harmony with natural human impulses. Among the American Indians, for example, boys were trained in active outdoor pursuits—just the sort of things boys like. But in our society children are pushed into studying technical subjects, which most do grudgingly.

Additional comments in the Manifesto give us clues to the bomber's inner thoughts:

The moral code of our society is so demanding that no one can think, feel and act in a completely moral way. For example we are not supposed to hate anyone, yet almost everyone hates somebody at some time or other, whether he admits it to himself or not.

In order to avoid feelings of guilt, they continually have to deceive themselves about their own motives and find moral explanations for feelings and actions that in reality have a non-moral origin. We use the term oversocialized to describe such people. Oversocialization can lead to low self-esteem, a sense of powerlessness, defeatism, guilt, etc. If a particular child is especially susceptible to such feelings, he ends by feeling ashamed of HIMSELF.

HARVARD

ALMOST EVERYONE WILL AGREE THAT WE LIVE IN A
DEEPLY TROUBLED SOCIETY.

UNABOMBER MANIFESTO

UNLIKE THESE PROOFS, THE PROOF GIVEN HERE IS
GROUP-THEORETIC, IN THE SENSE THAT THE ONLY
NON-GROUP-THEORETICAL CONCEPTS EMPLOYED ARE
OF AN ELEMENTARY NATURE.

T. J. KACZYNSKI
"ANOTHER PROOF OF WEDDERBURN'S THEOREM"

Off to Harvard at 16, Kaczynski managed to share a suite in preppy Eliot House with several fellow students, but he made little impression on them. "We had no interaction," said Michael Rohr, a former suite mate and now a philosophy professor at Rutgers. "I can't remember having a conversation with him."

From a purely academic standpoint, Kaczynski's professors couldn't help but be highly impressed by his brilliance. Robert Rippey, one of his math teachers, commented:

> What made him special was the way he thought. He'd take a mathematical proof and find different ways of doing it, without any instruction from me. It's the hallmark of a truly penetrating mind.

Despite (or because of) his genius in mathematics, Kaczynski sunk deeper into solitude. Increasingly isolated, he began bringing sandwiches from the dining room upstairs to his room, which soon began to smell of spoiled milk, rotting food, and foot powder. Suite mate Patrick McIntosh (now a Colorado astronomer) noted,

> In some places, the papers and such were a foot deep. That disturbed me, that someone could live in such filth. The worst part was when it began to smell. Maybe it was rancid milk.

Kaczynski began to develop the odd metronomic habit of rocking back and forth on a chair as he studied. (I do not know if this behavior was involuntary, but it reminds me of the American theoretical physicist Richard Feynman [1918–1988], who, while at MIT, fell into the habit of bothering his roommates with the absentminded drumming of his fingers and a tapping staccato on walls and garbage pails. See Appendix A.) Kaczynski frequently strode through the suite, saying nothing, and slammed his door to shut out his suite mates. If one of the suite mates approached him he would run to his room to avoid conversation.

Kaczynski continued to find it painful to make errors and compulsively corrected minor errors in others. He shut himself up in his bedroom for days at a time, and seemed incapable of sympathy, insight, or the simplest connections with people. When his brother would send him letters later in life, he would return them with grammatical corrections! Kaczynski graduated from Harvard with a mathematics degree in 1962, just days after he turned 20. Later in life, in a report for the 20th and 25th reunions of the Harvard Class of '62, he listed his address as:

T J Kaczynski
788 Banchat Pesh
Khadar Khel, Afghanistan[5]

No such place exists.

GRADUATE SCHOOL AND BEYOND

BY MEANS OF AN INTRICATE CONSTRUCTION, THE
AUTHOR PROVES THAT FOR ANY SET *A* ON THE UNIT
CIRCLE OF TYPE *F(SD)*, AND FOR ANY FUNCTION *T*
DEFINED ON *A* THAT DIFFERS FROM SOME FUNCTION
OF THE FIRST BAIRE CLASS AT MOST COUNTABLY
MANY POINTS, THERE EXISTS A CONTINUOUS
COMPLEX-VALUED FUNCTION *F* DEFINED IN *D*
HAVING *A* AS ITS SET OF CURVILINEAR
CONVERGENCE AND HAVING *T* AS
ITS BOUNDARY FUNCTION.

T. J. KACZYNSKI
*"BOUNDARY FUNCTIONS AND SETS OF CURVILINEAR
CONVERGENCE FOR CONTINUOUS FUNCTIONS."*

Like a diver plunging deeper and deeper into the mathematical ocean, Kaczynski devoted himself to writing about increasingly more abstract mathematical ideas. His dissertation, entitled "Boundary Functions," earned him a cash award from his department. It focused on functions (mathematical entities that assign to each element of one set at least one element of the same or another set). In particular Kaczynski was interested in functions as they relate to circles. His ideas had no apparent practical application but the papers are brilliant according to seasoned mathematicians. A typical sentence in his thesis reads

> If f is defined in H, then the set of curvilinear convergence of f is the set of all points $x \in X$ such that there exists some arc at x along which f approaches a limit.

Peter Duren, who sat on the committee that judged Kaczynski's doctoral thesis, noted,

> He was one of my best students, one of only four students who got A's in the class I taught. Kaczynski also took one other

class—Human Evolution. He earned Professor Frank Living-
stone's first A-plus in five years.

Kaczynski was a meticulous, top student who never made mistakes.
At graduate school, he wrote with a draftsman's hand and provided more
proofs than necessary. In his paper "On a Boundary Property of Continuous Functions," Kaczynski used uncharacteristic restraint when he
wrote, "One could go on listing such corollaries ad infinitum, but we
refrain." One wonders whether his frequent attention to backup proofs
might have something to do with the redundancy he built into his bombs,
providing more initiators than the devices required.

Kaczynski's research into the properties of functions of circles was by
all accounts brilliant, and when he sent his papers to journals for publications, he did so quietly, without telling his professors or classmates. When
his articles began appearing in respected mathematics journals, professors
and students were amazed. According to Joel Shapiro, a fellow student
now a mathematics professor,

> While most of us were just trying to learn to arrange logical
> statements into coherent arguments, Ted was quietly solving
> open problems and creating new mathematics. *It was as if he
> could write poetry while the rest of us were trying to learn grammar.*

Despite his brilliance as a thinker and mathematical researcher, his
attempts at teaching were a disaster. While at the University of California,
Berkeley, he taught at least five courses: Advanced Calculus for the Applied Sciences, Introduction to the Theory of Sets, General Topology, Function Spaces, and Number Systems. Student questionnaires suggest that his
lectures were "useless, right from the book." Other responses indicated
that he showed no concern for the students. "He absolutely refuses to
answer questions by completely ignoring the students."

There was one woman who tried to reach out to him only months
before he abandoned his efforts to build a life at Berkeley. Graduate
student Harriet Hungate took Kaczynski's topology class. (Topology is a
branch of mathematics that deals with properties of objects when they
undergo distortion. A lump of clay, for example, may be regarded as a
collection of physical points that can be deformed, say, into a ball or a long,
thin rod without changing topologically.) Harriet was impressed by Kaczynski. She considered him young, tall, and attractive "with all the outward manifestations of someone who would be very sociable."

It was a particularly emotional time in Hungate's life, and she needed someone with whom she could commiserate. She was in crisis, thinking of dropping out, and she decided to go to Kaczynski's office for a conference. She asked questions about the course, but then her emotions took over. She spilled out her fears and doubts, looking to her professor for warmth and reassurance.

His response scared her. Kaczynski did not acknowledge she had spoken.

"There was no reaction at all," Hungate said. "Usually there's a person in there who responds to what is said, but I looked in his eyes, and I saw no person there."

As an assistant professor in Berkeley's math department, Kaczynski seems to have slipped further and further away from the world around him. He suddenly resigned after teaching for two years, giving no hints as to his reasons:

> January 20, 1969
>
> Dear Professor Addison,
>
> This is to inform you that I am resigning at the end of this academic year. Thus I will not be returning in Fall, 1969.
>
> Sincerely yours,
> T. J. Kaczynski

That was it. No explanation. The department chair, John W. Addison, Jr., tried and failed to talk him into staying.

Between the time he left the university and went off into the woods, Kaczynski stayed mostly in his bedroom in his parent's home. He did nothing for more than a year, even when his parents urged him to get a job to give him something to do.

KACZYNSKI'S MATHEMATICAL WORK

LET K BE AN ALGEBRAIC SYSTEM WITH TWO BINARY OPERATIONS, SATISFYING (1) K IS AN ABELIAN GROUP UNDER ADDITION, (2) $K - \{0\}$ IS A GROUP UNDER MULTIPLICATION, AND (3) $x(y + z) = xy + xz$ FOR ALL $x, y, z \in K$. SUPPOSE THAT FOR SOME N, $0 = 1 + 1 + \ldots + 1$ (N TIMES). PROVE THAT, FOR ALL $x \in K$, $(-1)x = -x$.

T. J. KACZYNSKI
"DISTRIBUTIVITY AND $(-1)x = -x$"

Various mathematicians have said that Kaczynski's papers such as "Boundary Functions for Functions Defined in a Disk" and "On a Boundary Property of Continuous Functions" were cutting edge mathematics at the time. In order to help judge his work, I've acquired his papers and spread them out before me. Alas, despite some mathematical training, I cannot understand Kaczynski's work. Here are some of his erudite titles:

- Kaczynski, T. J. (1967) *Boundary Functions* (doctoral dissertation). Ann Arbor: University of Michigan. (This 80-page thesis won "best thesis of the year" in the math department at the University of Michigan.)
- Kaczynski, T. J. (1964) Another proof of Wedderburn's theorem. *American Mathematical Monthly.* 7(1): 652–653.
- Kaczynski, T. J. (1964) Distributivity and $(-1)x = -x$ (proposed problem). *American Mathematical Monthly.* 71: 689.
- Kaczynski, T. J. (1965) Boundary functions for functions defined in a disk. *Journal of Mathematics and Mechanics.* 14(4): 589–612.
- Kaczynski, T. J. (1965) Distributivity and $(-1)x = -x$. (problem and solution). *American Mathematical Monthly.* 72: 677–678.
- Kaczynski, T. J. (1966) On a boundary property of continuous functions. *Michigan Mathematics Journal.* 13: 313–320.
- Kaczynski, T. J. (1969) Note on a problem of Alan Sutcliffe. *Mathematics Magazine.* 41: 84–86.
- Kaczynski, T. J. (1969) The set of curvilinear convergence of a continuous function defined in the interior of a cube. *Proceedings of the American Mathematics Society.* 23: 323–327.
- Kaczynski, T. J. (1969) Boundary functions and sets of curvilinear convergence for continuous functions. *Transactions of the American Mathematics Society.* 141: 107–125.
- Kaczynski, T. J. (1969) Boundary functions for bounded harmonic functions. *Transactions of the American Mathematics Society.* 137: 203–209.

I shouldn't feel too bad though. There are very few who can fully appreciate his work. Professor Maxwell O'Reade, who was on Kaczynski's dissertation committee, noted, "I would guess that maybe 10 or 12 people in the country understood or appreciated it."

Just how amazing a find is it for only a handful of people to be able to understand mathematical papers such as Kaczynski's? To put this in perspective, consider that the body of mathematics has generally increased from ancient times, although this has not always been true. Mathematicians in Europe during the 1500s knew less than Grecian mathematicians at the time of Archimedes. However, since the 1500s humans have made tremendous excursions along the vast tapestry of mathematics. Today there are probably around 300,000 mathematical theorems proved each year.[6]

In the early 1900s, a great mathematician was expected to comprehend the whole of known mathematics. Mathematics was a shallow pool. Today the mathematical waters have grown so deep that a great mathematician can know only about 5 percent of the entire corpus. Even though his papers are considered important to those in Kaczynski's area of math, I wonder what will the future of mathematics be like as specialized mathematicians know more and more about less and less until they know everything about nothing? Had Kaczynski continued, what important contributions might he have made to mathematics?

PARENTS

IF S IS A GENERALIZED QUATERNION GROUP, THEN S CONTAINS A QUATERNION SUBGROUP GENERATED BY TWO ELEMENTS A AND B, BOTH OF ORDER 4, WHERE $BA = A^{-1}B$. NOW A^2 GENERATES A COMMUTATIVE FILED IN WHICH ONLY THE ROOTS OF THE EQUATION $X^2 = 1$ OR $(X + 1)(X - 1) = 0$ ARE ± 1.

T. J. KACZYNSKI
"ANOTHER PROOF OF WEDDERBURN'S THEOREM"

Some of the strange brains in this book, such as Bentham and Heaviside, have had problems relating to family members, particularly their fathers. Like Isaac Newton, who had a hatred of his parents, Kaczynski also had his own family animosities. In 1990, when his father died, he didn't attend his father's funeral. For some time Kaczynski had been decreasing his contact with family members, complaining he had developed a heart arrhythmia[7] made worse by dealing with them. To identify any "urgent and important" letters they might send, he asked them to draw a red line under the postage stamp on their letters. According to an April 22, 1986, issue of *Time* magazine, when they used a red line to mark the letter that told of his father's suicide, Kaczynski wrote back complaining that the message didn't merit a red line.[8] (His father had lung cancer, and when his condition deteriorated, he shot himself to death in his home.)

When his mother invited him to talk about the things that might have gone wrong in their relationship, Kaczynski wrote back a 17-page letter with offensive epithets accusing his parents of being "more interested in having a brilliant son than seeing that son happy and fulfilled."

From that point on he retreated to an even greater degree into the Montana woods and his books. But in his early days in the Chicago suburbs, he did not appear to be isolated from his family or have an embittered relationship with his parents. His father, Theodore R. Kaczynski, was a sociable, happy man who loved the outdoors, camping, and canoeing, and taught his sons to do the same. Friends said the sausage-factory owner was an atheist who liked to consider the big questions of life.

Kaczynski's mother, Wanda Kaczynski, was no overbearing tyrant driving her son beyond his capacity. Counselors and teachers—not his parents—suggested Kaczynski skip his junior year in high school, according to his chemistry, math, and physics instructor, Robert Rippey. The parents "just wanted him to have a good time, to do the things that interested him and not be bored," Rippey said.

Sadly, years later, Kaczynski would write raging letters from Montana blaming his mother for his social incompetence, calling her a "dog." However, neighbors of the Kaczynskis have recalled that both parents appeared to be loving to their children.

BROTHER

If there is a genetic or environmental component to strangeness and brilliance, we might briefly explore the life of Kaczynski's brother. Despite the seven-year difference in their ages, the Kaczynski brothers exhibited startling similarities. Both were quiet, smart, and introspective. Both went to Ivy League schools but didn't use their prestigious degrees for advancement or materialistic gain. They both sought solitude in remote parts of the country where they tried to survive in stark conditions and enjoy the natural world. Both brothers appear to have compelling notions about justice. Ted's may have led him to murder. David's led him to tell the FBI about his suspicions that Ted was the Unabomber.

In the 1980s, David retreated onto 30 acres of land among the mesquite and greasewood of the Chihuahuan desert. Lizards and diamondback rattlers were more frequent than humans. He lived at first in a four-foot hole dug in the ground, covered with tarps, until he built his own primitive cabin in the wilderness of the Christmas Mountains of West Texas where he was 20 miles from the nearest paved road. Like Kaczynski, David was resourceful. A strict vegetarian, he ate berries from cacti and skinned and fried large cactus pads in a skillet.

David revered his brother's passionate commitment to wilderness existence. Because his cabin had no running water, David sometimes showered at a bunkhouse maintained by the Terlingua Ranch. It was there that in 1983 he met Juan Sanchez, a Mexican farmhand who did maintenance work at the estates. David helped him secure a green card and suggested he write to Ted to get more advice on his immigration problems. That led to a seven-year correspondence. From November 1988 to November 1995, Ted wrote to Sanchez almost every month in neatly penned, grammatical Spanish. Ted often wrote of his poverty and isolation. With a mathematician's accuracy he described his finances down to the penny. "As to my poverty, I have $53.01 exactly, barely enough to stave off hunger this winter without eating rabbits." At Christmas time 1994, Ted sent Sanchez a carved wooden tube marked with a Latin motto, *Montana Semper Liberi* ("Mountain Men Are Always Free").

David was like his brother in many other ways, including his meticulousness. For example, once when David was hiking he picked up an old Indian flint knife. When he learned that it would be considered an artifact and wrong to have moved it, he hiked back 40 miles to return it to where he had found it. David also worried about the threat of technology and what it would do to humanity.

Despite these similarities, David was more personable than Ted, and he eventually integrated himself into society. In 1990, just before their father killed himself, David came north, cut his hair, married his high school girlfriend, and even wired his cabin for a computer. His wife teaches philosophy at Union College in Schenectady, New York, near where they live. For the past several years, David has been a counselor for troubled teens at a youth center in Albany.

David, who graduated from Columbia University in 1970 with a degree in English, tried a number of jobs. One was as a supervisor at the Addison, Illinois, factory Foam Cutting Engineers, a company that also employed his father. For awhile in 1978 one of his subordinates was Ted, who had left Montana briefly in the hope of earning some money. Ted went on a date with a female supervisor at the factory. After their relationship did not materialize, he responded by composing crude limericks about her and posting them around the plant. When David told him to stop, Ted pasted one onto the very machine his brother was operating. David quickly fired him. Soon after, Ted wrote the woman a letter in which he said he had considered doing harm to her. She was lucky, he told her, that he had decided not to. We do not know how serious the threat was, but the first Unabomb explosion took place a few months before his firing.

ROASTED SQUIRRELS

IF -1 IS A QUADRATIC RESIDUE, TAKE $R = 0$ AND CHOOSE T APPROPRIATELY. ASSUME -1 IS A NONRESIDUE. THEN ANY NONRESIDUE CAN BE WRITTEN IN THE FORM $-S^2(\text{MOD } Q)$ WITH $S \neq 0$.

T. J. KACZYNSKI
"ANOTHER PROOF OF WEDDERBURN'S THEOREM"

What was Kaczynski's life like as a hermit in the woods? For one thing, we know he made his own bread and candles, and his cabin was built out of plywood and rolled roofing paper. The cabin had two windows, each a foot square, but no electricity, plumbing, or phone. He grew parsnips, potatoes, carrots, and other vegetables, and he used a bucket for a toilet. For fertilizer he used his own waste, and constantly battled the deer that dug through the garden's chickenwire fence. Kaczynski learned to be a very quiet, very careful, and a very patient hunter. He hunted squirrels and rabbits and roasted them over a pit outside his cabin. His life was largely financed by his parents, who gave him a $1000 to $1500 a year in birthday and Christmas gifts, and he kept a careful record of any expenses.

David, his brother, said that Kaczynski expressed frequent concerns about germs, inflections, and other health matters. "This kind of worry about his health is a recurring theme," David said. Preoccupation with germs seems to be a recurring theme with the strange brains in this book. Notice the irony in Kaczynski's preoccupation considering that he chose to live in a rather nonantiseptic environment.

Unlike his brother, Ted Kaczynski never found an escape from his forest of isolation. While in seclusion, the emotions that built inside him came out in letters and acts directed against people at safe distances—just as the Unabomber soon would attack and kill at safe distances. For example, aside from his scalding letters to his mother, he wrote a letter to a service station owner calling him a "fat con man." A neighbor's unoccupied cabin was trashed after Kaczynski complained their snowmobiles disturbed him.

He complained in person only to the phone company. He could not operate their pay phones even though the instructions were written on the front. One of his early letters to the Montana State Commerce Department regarding the pay phones has the flavor of a mathematical proof:

Theodore J. Kaczynski
HCR 30 Box 27
Lincoln MT 59639

Montana State Commerce Department
Consumer Affairs Unit
1424 9th Avenue
Helena, MT 59601

Dear Sirs:
    I have a complaint about the Lincoln Telephone Company. . . . The problem is that some of the Lincoln Telephone Company's pay phones malfunction in such a way as to steal the caller's quarters. You put a quarter in and it gets jammed, or it doesn't register, and the coin release doesn't work, so that either you can't put the call through and your quarters are lost, or else the call does go through and you've put into the phone 25 cents or 50 cents more than the price of the call. This problem has persisted for several years . . .

EVIDENCE

HENCE $A^{-1} = A^3 = -A$, SO $BA = -AB$.

T. J. KACZYNSKI
"ANOTHER PROOF OF WEDDERBURN'S THEOREM"

When the FBI finally descended on Kaczynski's 10-by-12-foot cabin, they found much evidence: bomb parts stashed in old food cans, a list of corporate executives stored along with a San Francisco city map, and an obscure book that was quoted in the Unabomber's infamous manifesto. A Samsonite briefcase held Kaczynski's master's and doctoral degrees from the University of Michigan. An envelope in the cabin was marked "Autobiography."

The contents of Kaczynski's cabin were itemized in a 700-item list filed in U.S. District Court in Helena, Montana. It detailed the results of a meticulous court-authorized FBI search that began April 3, 1996, soon after Kaczynski was detained on a possession of explosives charge while authorities tried to piece together his possible involvement in a nearly 18-year bombing spree that killed three and injured at least 22. According to Berkeley campus police Captain William P. Foley, the FBI also found the names of approximately 25 University of California mathematics professors in the cabin. About half are retired or semiretired professors who were at Berkeley when Kaczynski taught there in the late 1960s.

Hundreds of books were found in his cabin, including *Les Miserables* by Victor Hugo, *Growing Up Absurd* by Paul Goodman, six books on Eastern mysticism, and *Comes the Comrade* (a novel about a Hungarian village under Nazis and then Soviets). Additionally, there was a *Holy Bible Dictionary Concordance* and *Asimov's Guide to the Bible*. (The presence of Asimov's skeptical commentary as well as Kaczynski's dislike of Catholicism suggests to me that he was not religious.) A copy of the novel *Ice Brothers*[9] by Sloan Wilson is notable because a gutted copy of this book was used to deliver the bomb to the 1980 victim, United Airlines president Percy Wood.

Another of Kaczynski's books, *Violence in America: A Historical and Comparative Perspective*, also provides evidence linking him to Unabomber attacks. This book, commissioned by a government task force in the 1960s, is quoted extensively in the Unabomber's manifesto, suggesting that investigators may have found not only the manifesto, but the literature used to produce it.

A number of cans—Calumet Baking Powder, Shoppers Value Quick Oats, Quaker Yellow Corn Meal, Quaker Oats, and Old Fashioned Quaker Oats—stored things having nothing to do with breakfast or baking. Inside were possible bomb-making accessories. One can labeled "Del Monte Whole Leaf Spinach" contained melted metal fragments that can be used for shrapnel in bombs. A "Tater Tot" box contained books and maps; a Tide detergent box, tools. Two jars contained ammonium nitrate, a bomb-making component.

There were dozens of chemicals in the cabin, and Kaczynski was an ardent labeler as evidenced by labels[10] on various receptacles: "Perfectly clean," "$BaSO4$ may be contaminated with a little $MgCl$ and smaller amount of $MgSO2$," and "Al compounds and other crap." A pipe bomb was found, as was an "improvised explosive device" in a cardboard box, and five guns (one was one a big-game rifle, and another was handmade).

On June 25, 1996, Theodore Kaczynski, age 54, pleaded innocent to a 10-count federal indictment accusing him in four of the 16 Unabomb attacks, including two fatal explosions 10 years apart. Kaczynski faced charges of transporting, mailing, and using bombs. At that time he could have faced the death penalty if convicted of any one of three charges relating to the fatal attacks that included a 1985 blast that killed computer rental store owner Hugh Scrutton and a 1995 explosion that killed timber industry lobbyist Gilbert Murray.

While in jail, Kaczynski was kept away from other prisoners, and although inmates are permitted visitors, sheriff's spokeswoman Sharon Telles said Kaczynski was not interested in seeing anyone. "He's being a civil, orderly prisoner—doing what's requested of him," she said. "He's just a rather reserved person."

Even before federal agents searched Kaczynski's mountain cabin for evidence, DNA tests of saliva found on two letters had already linked him to the Unabomber, according to court documents. In particular, forensics experts have concluded that the genetic content of saliva on a postage stamp for a Unabomber letter was similar to DNA found on a letter Kaczynski sent to his brother, David.

Working with documents that David Kaczynski provided, the FBI found 160 similarities between the 35,000-word Unabomber manifesto and a 1971 essay written by Kaczynski. For example, in the earlier essay Kaczynski wrote: "direct physical control of the emotions via electrodes and 'chemitrodes' inserted in the brain." The 1995 Unabomber manifesto includes this similar sentence: "It presumably would be impractical for all people to have electrodes inserted in their heads."[11] Both Kaczynski and the Unabomber spelled analyze as "analyse," used "wilfully" rather than willfully, and spelled license as "licence." Additionally, his letter stating "primitive people . . . may have had some elaborate process for making edible, as with certain other plants" has similarities with, "When primitive man needed food he knew how to find and prepare edible roots," which is from the Unabomber manifesto.

According to the September 21, 1996, *Mercury News* the FBI found that Kaczynski kept a day-to-day journal of his hermit lifestyle in the Montana woods that included entries about mailing bombs and describing their results. U.S. Attorney Robert Cleary said the journal contains "Mr. Kaczynski's admissions, detailed admissions, to each of the 16 Unabomb devices. In a number of instances, it is just 'I mailed that bomb'; 'I sent out that bomb.' Others expressed his desire to kill."

Could Kaczynski had been an obsessive–compulsive like others in this book? Certainly some of his behaviors point in this direction—his obsession with perfection and germs and the meticulous nature of his pristinely hand-crafted bombs. Interestingly, the FBI found a bottle of trazodone pills in his cabin when Kaczynski was arrested. This suggests that Kaczynski was aware of certain mental problems and that he was seeking treatment. Trazodone (brand name Desyrel) is an antidepressant. (The drug did not come from the local drugstore, because the druggist at the nearby Lincoln Pharmacy said he had filled no such prescription for Kaczynski.) Depression is a major theme of Unabomber's manifesto, which describes a world in which people can't achieve meaningful goals. In fact, the Unabomber describes the increased use of antidepressants as part of society's mind control attempts: "Instead of removing the conditions that make people depressed, modern society gives them antidepressant drugs."

Trazodone tends to be quite sedating, and is sometimes used as a sleeping aid or as an adjunct to other antidepressants in cases where they are having an adverse effect upon sleep. One infrequent but significant side effect in males is priapism: prolonged and painful erection of the penis that may require surgical intervention (e.g., injection of a vasoconstrictor) to resolve and prevent permanent damage.

Several people have reported heart arrhythmias as a result of taking trazodone, and one wonders if Kaczynski's claim that he had arrhythmias had any relationship to this drug. We do not know if trazodone had any effect on Kaczynski or could have influenced his killing spree. However, adverse behavioral reactions to trazodone have been reported, and these include drowsiness, fatigue, lethargy, retardation, lightheadedness, dizziness, difficulty in concentration, confusion, impaired memory, disorientation, excitement, agitation, anxiety, tension, nervousness, restlessness, insomnia, nightmares, anger, hostility, and, rarely, hypomania, visual distortions, hallucinations, delusions, and paranoia.

The Unabomber—Theodore J. Kaczynski—did not want to remain unknown despite the fact that he was a hermit who cleverly hid his identity. In an apparently arrogant manner, he left various clues for the world to ponder. The bombs were inside boxes fashioned of four kinds of wood and sent to forestry officials and intended victims named Wood. Like other strange brains in this book, the self-importance of his ideas seemed to be made more and more evident. "In order to get our message before the public with some chance of making a lasting impression, we've had to kill people," he wrote. He craved the publication of his Manifesto, which suggested supplementary readings. He strangely left hints that he was collective voice of an organization called "FC." Kaczynski's excessive shyness, concerns about germs, social isolation, and other extreme behaviors coupled with his mathematical genius make him an ideal candidate for this book.

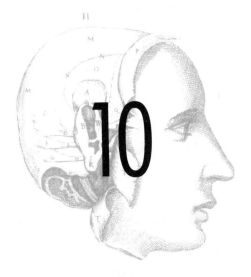

10

OBSESSION

INTRODUCTION

WHEN WE LEARN MORE ABOUT THE FUNCTION GONE
WRONG IN OBSESSIVE–COMPULSIVE DISORDER, WE
WILL ALSO LEARN MORE ABOUT THE MOST
MYSTERIOUS SECRETS OF THE NATURE OF MAN.

JUDITH RAPOPORT, M.D.
THE BOY WHO COULDN'T STOP WASHING

THIS IS PATENTLY ABSURD; BUT WHOEVER WISHES
TO BECOME A PHILOSOPHER MUST LEARN NOT TO BE
FRIGHTENED BY ABSURDITIES.

BERTRAND RUSSELL

Here we take a break from the biographies of obsessive–compulsive geniuses to explore obsessive–compulsive disorder itself. Did Nikola Tesla, Oliver Heaviside, and other geniuses exhibit creative greatness *because* of their obsessions and compulsions or in *spite* of them? After poring through biographies of geniuses, I believe that there is an interplay of factors. Eccentricity sets geniuses apart from ordinary people, forcing them to struggle and rise above their affliction. In some cases, as with bipolar (manic–depressive) artists, creativity seems to be linked with geniuses' mental illness. Clearly, many great individuals through history have exhibited what modern physicians would classify as obsessive–compulsive disorder, an emotionally crippling sickness that today afflicts around six million Americans. People tormented by this disease have little control over their unusual behaviors, which often include excessive cleaning, counting, washing, checking if doors are locked, and avoiding situations that set off these odd patterns. Most of the time, people with obsessive–compulsive disorder know that their behavior is illogical or self-destructive, but like someone with a strong addiction, they find it impossible to stop. For example, a person may spend five hours a day washing himself and still feel dirty. Another sufferer must check the door hundreds of times a day to make sure it is locked. Some pluck out every hair on their heads. Children may play endlessly with strings or pick up objects with their elbows to avoid getting their hands dirty. These children usually continue to suffer the same symptoms as adults if untreated.

Here are some obsessive–compulsive disorder facts:

- Half of all obsessive–compulsive disorder suffering starts in childhood; it usually occurs in boys.
- If obsessive–compulsive disorder starts later in life, there is a 50:50 mix of men and women.
- 85 percent of people with obsessive–compulsive disorder have at one point suffered from excessive washing.
- A high proportion of obsessive–compulsive disorder sufferers have had minor tics or twitches of their face or hands.

Nikola Tesla's obsession with germs, counting food items, or insisting things come in multiples of three are all examples of obsessive–compulsive disorder. Howard Hughes, the wealthy inventor, also had a preoccupation with germs that, ironically, led to a bizarre life of filth and neglect. Like inventors Tesla and Heaviside, Hughes preferred darkened, shuttered rooms sealed off from the rest of the world. Hughes required his servants to bring him food and other items while wearing special tissue pads to avoid touching anything that he might touch. The list of brilliant people with obsessive–compulsive disorder is endless. Recall our discussions in Chapter 3 of Samuel Johnson (1709–1784), a genius clearly afflicted with obsessive–compulsive disorder. As just one example of his behavioral disorder, this scholar, poet, and witty playwright was forced to go through elaborate rituals whenever entering doorways.

A FEW EXAMPLES IN THE NORMAL POPULATION

People afflicted with obsessive–compulsive disorder perform endless rituals that dominate each of their days. They usually recognize the problem as soon as it starts and therefore often keep their embarrassing disease hidden. Judith Rapoport, M.D., in her impressive book *The Boy Who Couldn't Stop Washing*, describes many interesting cases, a few of which are outlined below to give you a flavor of the range of possible symptoms:

- *Circle Boy*—A two-year-old boy feels compelled to walk in circles around a manhole as if his brain were caught in an infinite programming loop. At the age of ten, the same boy is removed from school because of his bizarre compulsion to draw O's.
- *Saliva Child*—A six-year-old boy has to crouch down and touch the ground whenever he swallows saliva. Any wastage or loss of saliva is unthinkable. Later in life, the boy must blink his eyes or touch his shoulders to his chin whenever he swallows his saliva. He says, "I was frustrated because I couldn't stop my compulsions. I don't know why. I had no reason." The same boy is compelled to rip toilet paper into tiny pieces. In addition, he is compelled continually to touch the ends of his thumbs to the opening in water faucets.
- *Sticky Boy*—A 14-year-old boy is revolted by sticky things. Honey is the ultimate horror. He says, "Stickiness is terrible. It is some kind of disease. It is like nothing you can understand."
- *Fence Boy*—A seven-year-old boy can't go through the doorway in his fence. His sister must push him through.
- *Sugar Queen*—Over a period of 20 years, a middle-aged woman sprinkles confectioner's sugar over every horizontal surface in her house, including the floors and furniture. Her compulsion started rather suddenly when she felt the overriding curiosity to know how it would be to see and feel the sugar.

- *Letter Girl*—A seven-year-old girl's handwashing compulsion evolves into a compulsion to fill in the space in every letter with a closed area such as an o, p, a, q, or d. (This reminds me of some of the strange compulsions Heaviside had regarding the letters of his name.)
- *Sinning Girl*—A sixth-grade girl feels she is offending God whenever she touches the table or touches her hands. She volunteers to be punished at her school because her breathing is too loud. She goes to a priest who cautions her to pray no more than five minutes a day. Another priest *adds* prayers to her daily schedule, which makes her feel better for a while. This form of obsessive–compulsive disorder is known as *scrupulosity* and it manifests itself by excessive praying, unreasonable doubting, and the continued feeling that one is sinning. Many afflicted with scrupulosity wash their hands excessively.
- *Rat Man*—Sigmund Freud describes a "rat man" tormented by obsessive thoughts of rats eating his anus.

NUMERICAL OBSESSIVE–COMPULSIVE DISORDER

Obsessive–compulsive disorder involving numbers is particularly fascinating. Nikola Tesla had "arithromania" or "numerical obsessive–compulsive disorder." Recall that he demanded precisely 18 clean towels each day. If asked why, Tesla provided no explanation. Table accoutrements and towels were not the only items he demanded come in multiples of three. For example, he often felt compelled to walk around the block three times. Recall that he always counted his steps while walking. He chose room number 207 in the Alta Vista Hotel, because 207 is divisible by 3.

Does numerical obsessive–compulsive disorder involve *particular* numbers? Are obsessions with odd numbers more likely than even? Do obsessions ever involve numbers larger than ten? To better understand numerical obsessive–compulsive disorder, I pored through many case histories and created a list sorted by the number with which the person was obsessed:

- 1—No cases found.
- 2—No cases found.
- 3—A 13-year-old girl (see "9") is compelled to knock three times on the edge of the window and three times on a nearby door before

unlocking the door. (Recall that Tesla's love for the number three was discussed in Chapter 1.)

- 4—Case 1: An 11-year-old boy's life is ruined because the number 4 dominates his life.
Case 2: A teenage boy must have everything in 4s and avoids 6s. (He also has the compulsion to see the bottoms of his feet whenever he looks at the clock in his room.)
- 5—No cases found.
- 6—Case 1: A college boy avoids repeating any actions 6, 13, 60, 66, or 130 times. Multidigit numbers (such as 43 or 33) adding up to 6, 13, or 130 must be avoided.
Case 2: A teenage boy begins his day normally when suddenly the only thing on his mind are the repeating numbers "6, 6, 6, 6" or "8, 8, 8, 8." He reports, "I had no control over these numbers, they had a mind of their own—*my* mind."
- 7—The 11-year-old boy listed under "4" suddenly switches to a hepta-philiac when, after a brain operation, he has the very time-consuming compulsion to touch everything 7 times and ask for everything in 7s. He swallows 7 times. (His heptaphilia is cured by Anafranil, a drug discussed in greater detail in the "Cures" section.)
- 8—Case 1: A 12-year-old boy is compelled to turn around exactly 8 times in a coat room in order to calm himself.
Case 2: A boy in the shower strokes the right side of his head 8 times, applies shampoo, then strokes another 8 times, rinses 8 times, and strokes 8 times again. He repeats the process for the left side.
- 9—A 13-year-old girl must lift her feet and tap 9 times on the edge of her bed before climbing into bed.
- 22—An 18-year-old boy is compelled to count to 22 over and over again. He taps on the wall 22 times or in multiples of 22. He walks through doorways 22 times and gets in and out of his chair 22 times. The boy becomes addicted to drugs which have interesting effects on his 22-ness. For example, while on amphetamines and cocaine, his 22 tapping increases to the point where all his time is spent tapping out 22 all over his walls. LSD makes the ritual completely disappear.
- 50—A seven-year-old girl ("Letter Girl" in previous section) must count to 50 in between reading or writing each word. This time-consuming ritual makes her an extremely slow reader in the second grade.
- 13, 60, 66, 130—See college boy listed under "6."

- 100—The 13-year-old girl listed under "9" must also count to 100 after brushing her teeth.

CURES

The cause of obsessive–compulsive disorder seems to be biological; in particular, obsessive–compulsive disorder results from an imbalance in the brain's chemistry involving the naturally occurring brain chemical serotonin. In the past, extreme treatments for obsessive–compulsive disorder involved removal of part of the brain's frontal lobes. For example, in the early 1960s, there was the case of a 60-year-old man who suddenly developed a compulsion to pick up small pieces of trash and store them in his house. As his rooms filled from floor to ceiling with bits of paper, he and his wife decided to seek medical attention. Physicians cut the connections between his frontal lobes, curing his obsessive–compulsive disorder. Unfortunately such a drastic operation led to personality changes including frequent pinching of women and urinating on the street.

In the 1980s there was the case of a man who had a severe hand-washing compulsion that eventually led him to attempt suicide. He shot a bullet into his head, but survived and never had obsessive–compulsive disorder again. Evidently the destruction of part of his frontal lobes cured him, and he attended college and now leads a normal life. Psychiatrist Leslie Solyom of Vancouver's Shaughnessy Hospital reported the case in *Physician's Weekly* and indicated the bullet achieved the same result as a surgical procedure called a leucotomy—a less radical version of a lobotomy—which is occasionally attempted as a last resort in treating patients with intractable obsessive–compulsive disorders. The leucotomy entails removing part of the left lobe of the brain. "He hit himself in exactly the same spot a surgeon would have in a leucotomy," Solyom said. "A millimeter higher or lower would have killed him." *Physician's Weekly* wryly called the case "successful radical surgery." But Solyom was quick to warn, "I wouldn't advocate people to shoot themselves to solve their psychiatric problems."

One of the modern cures for obsessive–compulsive disorder is the drug Anafranil (clomipramine), which affects the metabolism of serotonin in the brain. Other drugs such as fluoxetine (Prozac) and fluvoxamine (Luvox) also are useful.[1] The success with these drugs leads many researchers to believe that obsessive–compulsive disorder has a physical basis just like manic depression or epilepsy. LSD is another drug that affects serotonin levels and appears to "cure" obsessive–compulsive disorder. (The use of LSD outside of the laboratory may be dangerous. Mood shifts, time and space distortions, and impulsive or aggressive behavior are complications possibly hazardous to an individual who takes the drug.) Medical imaging studies suggest that obsessive–compulsive disorder is caused by an abnormality in a part of the brain known as the basal ganglia, which is buried deep within the brain and in the frontal lobes. In particular, one portion of the basal ganglia called the caudate nucleus appears to behave differently in people with obsessive–compulsive disorder. People with Tourette's syndrome and Parkinson's disease also have abnormalities in these areas.

Evidence continues to mount for obsessive–compulsive disorder's biological basis. For example, it appears to have a genetic component and to run in families. Some obsessive–compulsive disorder starts after a first epileptic seizure, as discussed in Chapter 16. The fact that many obsessive–compulsive disorder sufferers have movement disorders, such as facial tics at one point during the course of their disease, further suggests the biological cause for this disorder. Amphetamines make the disease worse probably because they affect the dopamine system, which acts against the serotonin system. As mentioned, LSD increases serotonin levels and reduces obsessive–compulsive disorder symptoms.

HAIR-PULLING WOMEN

One common form of obsessive–compulsive disorder, called trichotillomania, involves the compulsion to pull out hair, particularly eyebrows, eyelashes, and scalp hair. I do not know if individuals with this compulsion also pull out their pubic hair. Some of the hair-pullers rip out hair and eat the roots. Some pull out hair in an orderly manner. For example, if four hairs are pulled from the right side of the head, four would be pulled from the left.

Trichotillomania always begins in the teenage years and appears only in women.[2] Trichotillomaniacs usually have no other obsessive–compulsive disorder symptoms. After treatment with Anafranil, many women are able to resist the compulsion to pull their hair out.

Dr. Judith Rapoport has studied dozens of trichotillomaniacs, and she suggests that the behavior is a primate grooming ritual gone wild. She writes, "Talking to these women, all so amazingly alike, gave me the eerie sense that a primitive behavior pattern has come loose. An innate atavistic urge to groom, to preen that can't be suppressed."

Animals in the wild frequently groom themselves and each other. For example, rats spend about a third of their waking hours grooming, working their tongues from head to tail. Grooming serves several functions. It increases during times of stress and regulates temperature through the evaporation of saliva. It also influences sexual behavior, since there are different odors in male and female saliva. Grooming also removes parasites and cleans the body.

Researchers have found that a rat's grooming behavior can be turned on and off simply by injecting certain hormones or chemicals or destroying certain areas of the brain. In addition, rats with obsessive–compulsive disorder exist. These supergroomers or "barber rats" clip off their cagemates' hair and whiskers. Not much is known about barber rats except that do their supergrooming only in the dark. No one knows if supergrooming rats exist uncaged in the wild. In a manner reminiscent of supergrooming rats, some boys with obsessive–compulsive disorder will excessively lick their hands for hours.

Trichotillomania-like behavior has also been observed in monkeys, horses, sheep, cats, dogs, birds, and possibly reptiles. In all cases, the causes are unknown, and treatment is difficult. It is estimated that around 2 percent of the population have trichotillomania—not as rare as you might have expected. Trichotillomania appears to differ from other forms of obsessive–compulsive disorder in that more pleasure is experienced during the behavior. The current most effective human treatment is a combination of behavior avoidance therapy together with either Anafranil or Prozac, although full remission is rare.

Could obsessive–compulsive disorder stem from an overdeveloped cleaning program in some individuals? Almost all mammals engage in grooming, licking, scratching. Monkeys systematically search their fur, picking out pieces of dead skin and insects. Rats continually clean their nests. Maybe obsessive–compulsive disorder sufferers are running these same ancient programs but just at a higher rate than normal people.

COMPULSIVE HOARDING

Like rats, birds, and other animals that gather bits of twigs, string, and paper to build nests, some people with obsessive–compulsive disorder save virtually everything that they come in contact with: pieces of paper, trash, tissues, newspapers, old TV guides, shampoo bottles, calendars, and all the mail they have received. The rooms in their houses are junk heaps overflowing with empty bottles, catalogs, old clothes. It's as if some primitive nesting behavior has gone haywire.

As one example, a 24-year-old woman told Rapoport:

> There is a narrow path from my bedroom door to my bed and even there I walk on mail and newspapers. Aside from this path, there is no other place for me to walk, not even to my clothes closet unless I clear the way. I am ruining my life with this horrendous mess! I won't accept any dates; I don't see any friends, because they would have to come and see this. I don't know why I do it.

I have scoured the popular and scientific literature for advanced cases of hoarding and have uncovered two particularly fascinating examples. The first involves a man in Long Beach, California. He had been paying his rent regularly with five and ten dollar bills, so when he failed to appear with the rent in April 1975, his landlady became concerned. Entering his apartment, she found a strange rock collection but not a trace of the man.

The landlady looked around the apartment with growing apprehension. From floor to ceiling were neatly wrapped 100-pound packages of stones and concrete slabs. Even the closets and bathtub were completely filled. The 30 tons of rock buckled the floor. Like a stone pyramid in Egypt, there were various subterranean tunnels permitting access where he slept and to the toilet and couch. [Later in this book, I will discuss other more famous "burrowers," such as the great English physician William Harvey (Chapter 15) and Duke Bentinck-Scott (Chapter 15).]

As the landlady went closer to the rocks she noticed something even more weird. She stooped down to examine one of the packages which appeared to have words written on it. She slowly looked from one package to another and found that all had two words on them: "Me" and "Austria."

The man later resided at nearby motels, leaving similar packages of rubble in each motel. No one knows why the man gathered rocks. His current location is a mystery, and I do not know if he still collects rocks. Perhaps he has gone underground.

THE PACKRATS OF FIFTH AVENUE

My favorite example of excessive hoarding is the fascinating case of the Collyer brothers who lived over a half-century ago. They lived in a four-story Manhattan townhouse crammed with junk beyond imagination.

Over the years there had been a number of calls to police about the strange goings-on at the home at 2078 Fifth Avenue in New York City. On

March 21, 1947, police received a call reporting that there was a dead body in the house, and they responded immediately.

When the police arrived at the brownstone building, they found the doors locked and the lower windows tightly shuttered and boarded up. In the past when the police had knocked on the door, the men answered. This time was different. The policed decided it was time to chop through the front door to make sure no one was injured. To their surprise, the police found the hallway completely blocked by trash.

What should they do next? One enterprising young officer offered to climb a ladder and enter through a second-story window, and in a few minutes he climbed into the house. Slowly he made his way through mounds of trash. What was that in the corner? An old man's corpse sitting on the floor in a torn bathrobe. It was one of the brothers inconspicuously sheltered in a cave of debris. This 65-year-old elder of the brothers, once an admiralty lawyer, had become paralyzed, blind, and had not left the house for years. He had relied on his brother, who was four years younger, to provide him with all he needed.

But where was the younger brother? This once-friendly engineer and concert pianist was not in the house. Or so it seemed.

How did it all start? The brothers had lived in their house on Fifth Avenue since birth. Their mother raised them with gentle care. Their father had been a wealthy Manhattan obstetrician. When their mother died in 1933, the two brothers were emotionally devastated and never seemed to recover. They gradually became more and more withdrawn, never venturing far from home. The elder brother became blind in 1933 and paralyzed in 1940 but refused to see a doctor. The younger brother did his best to treat him by cramming 100 oranges into his older brother each week. When the older brother asked the younger about the efficacy of this diet, the younger's enigmatic response was, "Remember, we are the sons of a doctor."

The brothers soon stopped paying utility bills and never cared when their heat, electricity, and water were terminated. They cooked on kerosene stoves, carried water in buckets from a park four blocks away, and roamed the streets at night searching for food and supplies. This description should not suggest that they became subhuman monsters. On the contrary, the few people that came in contact with them found them polite, cultured, and courteous. They became the stuff of rumors and legends, and some neighbors suspected they lived like millionaires in opulent splendor within the shuttered windows of their home. It turned out that they had $100,000 in the bank, but they lived like vermin.

Obsession

In addition to the classic hoarding behavior of people with obsessive–compulsive disorder the brothers also had unreasonable fears—the foremost being burglary. As a result, they built horrendous barricades and booby-traps of junk in front of windows and doors. Secret, narrow mazes through the trash would allow the younger brother to traverse the house like a mouse in an underground burrow. Experts speculate that the younger brother saved the newspapers for the elder in case he would one day regain his sight.

Let's return to the story of the missing younger brother. After discovering the elder brother's body, police and sanitation workers began to sift through the wreckage in the home like archaeologists searching through the strata of time. After three weeks, police removed 120 tons of junk including:

- 14 pianos
- The dismantled components of a Ford model T
- 3000 books
- A seven-foot segment of a tree
- Old toys
- Tickets to a 1905 church gathering
- Pictures of pinup girls from 1910
- Several guns and swords
- Loads of unopened mail
- Various sewing machines and dressmaker's dummies
- One human corpse

On April 8, while searching through an eclectic montage of rubbish on the second floor, officials discovered a body. It was the younger brother squashed beneath a prodigious tower of newspaper. Apparently, while he was crawling through one of the tunnels to bring food to his elder brother, the younger triggered one of his own booby-traps and was crushed by a tower of trash. The apparent cause of death was suffocation. His immobilized body was gnawed by rats and was located just feet from where police had found his elder brother a few weeks earlier. Most likely, the elder had been waiting for the younger to bring him food and had finally starved to death.

The brothers and the rock-collecting man are examples of obsessive–compulsive disorder in relatively untalented individuals. But perhaps the most famous and creative hoarder of all times was the painter Pablo

Picasso (1881–1973). Since he probably collected sundry items for his art (much like an inventor collects gadgets for later use), Picasso's behavior is probably not sufficiently bizarre to invoke the diagnosis of obsessive–compulsive disorder, but his studios and homes were stuffed solid with junk. In 1973, Alden Whitman reported in Picasso's *New York Times* obituary:

> All his studios and homes—even the 18-room rambling La California at Cannes—were crammed and cluttered with junk: pebbles, rocks, pieces of glass, a hollow elephant's foot, a bird cage, African drums, wooden crocodiles, parts of old bicycles, ancient newspapers, broken crockery, bull-fight posters, old hats, and weird ceramics.

According to Picasso's mistress, Françoise Gilot, his bedroom was crammed with junk. She once wrote,

> At the far end was a high Louis XIII secretary and, along the left-hand wall, a chest of the same period, both completely covered with papers, books, magazines, and mail that Pablo hadn't answered and never would, drawings piled up helter-skelter, and packages of cigarettes. Above the bed was a naked electric light bulb. Behind the bed were drawings Pablo was particularly fond of, attached by clothespins to nails driven into the wall.

BIBLIOMANIA

I do not know if many psychiatrists consider "bibliomania" to be an obsessive–compulsive disorder, but the lengths to which some individuals will go to handle, possess, and accumulate books seems to me to be a form of madness of the obsessive–compulsive disorder variety.

Some bibliophiles go so far as to destroy their families and move out of their homes to accommodate their growing collection of books. For example, in the 1700s, bibliophile Thomas Rawlinson crammed his room at Gray's Inn so full of books that he had to sleep in the foyer. He finally moved into a large mansion and filled it to overflowing. When he died at aged 44, there was essentially no room to sit in the house.

M. Boulard, a Parisian bibliomaniac, indiscriminately bought 600,000 books. They were stacked in cupboards, attics, cellars. So heavy were the books that the house began to collapse like a pile of toothpicks. Boulard bought six more houses and filled them entirely with books.

Richard Heber, born in 1774, was yet another bibliomaniac. His book-buying exploits soared to new heights: It seemed he wanted to own multiple copies of every book ever published. He believed that every gentleman needed at least three copies of a book, one for his own reading, one to lend to friends, and one for his country house. Author Holdbrook Jackson describes Heber:

> A bibliomaniac if ever there was one. A bibliomaniac in the most unpleasant sense of the word; no confirmed drunkard, no incurable opium-eater had less self-control; to see a book was to desire it, to desire it was to possess it; the great and strong passion of his life was to amass such a library as no individual before him had ever amassed. His collection was omnigenous, and he never ceased to accumulate books of all kinds, buying them by all methods, in all places, at all times.

Heber died in 1833. Thomas Dibdin was the first to break into the home and view its contents. He writes,

> I looked around me in amazement. I had never seen rooms, cupboards, passages, and corridors, so choked, so suffocated with books. Treble rows were here, double rows were there. Hundreds of slim quartos—several upon each other—were longitudinally placed over thin and stunted duodecimos, reaching from one extremity of a shelf to another. Up to the very ceiling the piles of volumes extended; while the floor was strewn with them, in loose and numerous heaps. When I looked on all this, and thought what might be at [his other homes] it was difficult to describe my emotions.

Officials soon found three houses in England crammed with Heber's books. Other equally stuffed houses were located in Antwerp, Ghent, Paris, and Brussels. No one knows if he had other hidden warehouses of books scattered around Europe.

Finally, the greatest bibliomaniac of all time was Sir Thomas Phillipps, who lived a miserly life in order to devote his life to his one obsession. Born in 1792, he gradually acquired 100,000 books and 60,000 manuscripts, which at the time was more than all the libraries of Cambridge University.

Phillipps purchased some of his manuscripts at a sale of Heber's manuscripts in 1836.

Phillipps was a classic obsessive–compulsive disorder individual. He never threw away a scrap of paper, hoarding household bills and copies of correspondences. Although he collected priceless manuscripts, he also acquired old records thrown out by the government department and cartloads of waste paper on the way to be pulped.

In order to catalog his ever-growing collection, Phillipps enlisted the help of his wife and three daughters, who worked like slaves to list manuscripts. When his dining room became filled to overflowing, Phillipps simply locked it up, leaving his wife and three daughters to make do with three poorly furnished bedrooms and a sitting room. Lady Phillipps' dressing-table was the only area in the bedroom that remained free of books. After a few years of living like a rat in hell, she could no longer endure the never-ending inflow of books. She became a drug addict as the only means of escape and died at the age of 37.

After the death of his wife, Phillipps began searching for a replacement wife with more money to help him finance his collection obsession. He eventually settled for a clergyman's daughter and continued collecting with renewed vigor. "I wish to have one copy of every book in the world!!!!!" he wrote to a friend. He ate and slept among his books. His new wife eventually complained of rats, and had a nervous breakdown. Phillipps only watched as she was carted away to a cheap boarding house.

Phillipps died at the age of 80. Sifting through his books took several generations. Sales of his library's valuable contents continued into the 20th century as great treasures were gradually unearthed.

DID NIKOLA TESLA REALLY HAVE OBSESSIVE-COMPULSIVE DISORDER?

The Yale–Brown obsessive–compulsive disorder scale developed by Dr. Wayne Goodman lists dozens of symptoms displayed by individuals with obsessive–compulsive disorder. If you exhibit any of these, you may have obsessive–compulsive disorder, but don't diagnose yourself; see a specialist. Thoughts or habits are considered obsessions if they cannot be stopped and/or interfere with daily life. I've excerpted those symptoms from Dr. Goodman's list that apply to Tesla:

- Fear that something terrible might happen
- Concern with dirt or germs
- Intrusive (neutral) images, for example images of a cat

- Having to count to a certain number
- Excessive demand that others submit to your way of doing things

Oliver Heaviside appeared to have this disorder but perhaps to a lesser extent than Tesla. Certainly his W.O.R.M. title, preoccupation with the letters in his name, excessive writing, thermophilia, and fear of strangers suggest obsessive–compulsive disorder.

Here are some other possible symptoms of obsessive–compulsive disorder:

- Being overly judgmental of yourself or others
- Having illogical fears
- Obsession with a need for symmetry, exactness, or order
- Checking doors, locks, etc.
- Repeating rituals
- Hoarding and collecting

OBSESSIVE–COMPULSIVE DISORDER IN CYBERSPACE

After doing extensive research on obsessive–compulsive disorder, I formulated several questions that none of the literature seems to address. I therefore asked the following questions in the Usenet computer news-groups sci.psychology and alt.support.ocd. These are electronic bulletin boards that are part of a large, worldwide network of interconnected computers called Usenet. (The computers exchange news articles with each other on a voluntary basis.) I received very few responses, but I hope to fill in the answers in a future book.

1. Does obsessive–compulsive behavior always occur in an individ-ual's *simulations* of reality? For example, would a little girl who obsessively collected trash necessarily force that behavior on dolls in a make-believe dollhouse world? In virtual reality (VR) worlds of the future, would the obsessive–compulsive disorder of a partic-ipant occur in the virtual world in which he/she participated? (With computer-generated virtual realities, users don goggles and special gloves to enter lifelike worlds.) I predict that VR will be helpful in ameliorating some forms of obsessive–compulsive dis-order because VR can be used gradually and controllably to expose individuals to objects of fear. Consider an acrophobic who can safely peer over the side of a virtual building. By gradually expos-

ing people with obsessive–compulsive disorder to feared or "trigger" situations in cyberspace, therapists may be able to reduce anxiety. Already, some therapists have had success by introducing an individual to the very thing that triggers an obsessive–compulsive disorder attack and subsequently trying to reduce rituals. For example, a person with obsessive–compulsive disorder fearing dirt and germs might be made to hold his hands in toilet water for hours. This is called "exposure with response prevention." In some cases, this terribly unpleasant experience has desensitized the person with obsessive–compulsive disorder. (Recent research by Jeffrey Schwartz and his colleagues at the UCLA School of Medicine suggests that behavioral therapy for obsessive–compulsive disorder changes not only the behavior of people but also their brain chemistries.)

2. It appears that a significantly large number of genius scientists have had obsessive–compulsive disorder. Are there any published studies suggesting a correlation between obsessive–compulsive disorder and either intelligence, genius, or creativity? How many other famous cases of geniuses with obsessive–compulsive disorder exist?

3. Do people with obsessive–compulsive disorder exhibit obsessive–compulsive behavior in their dreams, or are they free of such behavior in dreams? The one response I received from an individual with obsessive–compulsive disorder is an emphatic "yes." He exhibits the disorder behavior while asleep and dreaming.

4. Is the incidence of "numerical obsessive–compulsive disorder" lower in societies with less emphasis on numbers or in preliterate societies?

5. Are there any known cases of individuals with the same obsessive–compulsive disorder who have married and collaborated on obsessive activities such as hoarding? Or has obsessive–compulsive disorder always been a solo, noncollaborative disorder? Do hoarders only hoard their own garbage? It turns out that the answer to the last question is "No." In 1988, one California family not only collected their own garbage, but also took home other people's garbage. The family was soon evicted from their bungalow because they would not clear out 25 tons of rat-ridden garbage. Only three years earlier, they had been evicted from another home for the same reason.

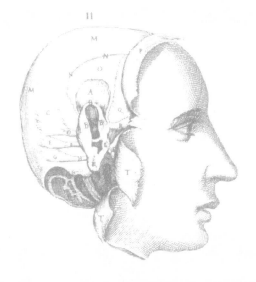

CURIOSITY SMORGASBORD

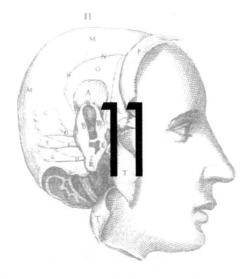

THE BRAIN SHELTER

"THE INCUBUS OF LEGS AND CHELAE AND VITAL
ORGANS WILL BE REMOVED. THE FUTURE KALDANE
WILL BE NOTHING BUT A GREAT BRAIN. DEAF,
DUMB, AND BLIND, IT WILL LIE SEALED IN ITS
BURIED VAULT FAR BENEATH THE SURFACE OF
BARSOOM—JUST A GREAT, WONDERFUL, BEAUTIFUL
BRAIN WITH NOTHING TO DISTRACT IT FROM
ETERNAL THOUGHT."
"YOU MEAN IT WILL JUST LIE THERE AND THINK?"
CRIED TARA OF HELIUM.
"JUST THAT!" HE EXCLAIMED. "COULD AUGHT BE
MORE WONDERFUL?"

EDGAR RICE BURROUGHS
THE CHESSMEN OF MARS

THE LINE BETWEEN SCIENTIFIC GENIUS AND
OBSESSIVE FANATICISM IS A FINE ONE.

THE BRAIN THAT WOULDN'T DIE (SCIENCE-FICTION
MOVIE)

E ver wonder where the world's largest collection of brains resides? Remember, at the book's outset, I had you imagining yourself the curator of a futuristic museum of brains. Perhaps you considered this idea bizarre or unlikely, but in England there *does* exist a huge brain shelter that holds the key to the enigmas of human behavior.

In a bucolic section of southern England 8000 wrinkled brains float in formaldehyde-filled jars. These brains, existing in obscurity for 40 years, are part of the largest brain collection in the world. The collection was established at the Runwell Psychiatric Hospital in 1950 by the late neuro-pathologist Nick Corsellis, who was interested in various forms of mental disease. Who would guess that this macabre storeroom of brains, deep inside an underground World War II air raid shelter, could offer today's scientists clues to brain dysfunctions caused by Alzheimer's disease, multiple sclerosis, schizophrenia, senile dementia, and boxing injuries? In fact, the collection contains all examples of mental disease, from Parkinson's to rare neurological conditions such as Pick's disease. Many brains in the collection were donated by patients or their relatives. The collection grew steadily in the 1950s, and the unused air raid shelter proved the largest and most suitable storage area. Many of the brains were housed helter-skelter in this old concrete bastion.

Open two jars in the shelter's collection, reach in, and remove two brains. Hold a brain in each of your hands. Can you see any differences? The brains of people with Alzheimer's appear shrunken and almost walnut-like. There is also a massive loss of nerve cells and white matter. Pry open a brain from someone who suffered from schizophrenia, and you will often seen abnormalities in the fluid-filled ventricles, particularly those affecting one part of the left hemisphere. It's hard to believe that only a decade ago schizophrenia was considered a functional disorder rather than the result of a brain tissue abnormality.

Dr. Clive Burton is the curator of the brain shelter, also called the Corsellis Collection after its founder Nick Corsellis. "The Corsellis Collection is larger by far than the two major brain banks in the United States and dwarfs all the other UK collections put together," Burton says.

The brain shelter has already proven useful for scientists studying brain disorders. For example, researchers examining the brains of 15 former British boxing champions found a definite link between the sport and specific brain damage. This information led to many important changes in boxing; for example, the number of rounds was reduced from 15 to 12 in some world championship fights, and headgear is now used in amateur fights.

In order to help researchers around the world, the brain shelter offers brain tissue samples of individuals who suffered from a range of disorders. A large database of patients' notes has been placed on the Internet, and wafer-thin samples of brain tissue preserved on slides can be mailed for a small fee. After spending two years cataloguing the specimens, Britain's medical research council is anxious to place the collection on a permanent and secure footing.

Today, local scientists are most interested in the possibility that criminality is an inherited neurological disorder. For example, Dr. Burton is currently studying aggression in the criminally insane on behalf of one of Britain's prison hospitals. He believes that a significant amount of criminal behavior is linked to structural and genetic brain damage. This is a very controversial subject with which many people disagree.

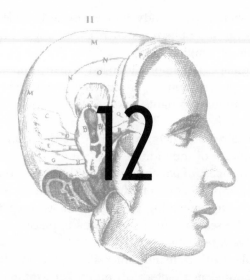

WHERE ON EARTH IS EINSTEIN'S BRAIN?

MOTHER WAS SHOCKED AT THE SIGHT OF THE BACK
OF ALBERT EINSTEIN'S HEAD, WHICH WAS
EXTREMELY LARGE AND ANGULAR, AND SHE FEARED
SHE HAD GIVEN BIRTH TO A DEFORMED CHILD.
MAJA EINSTEIN (ALBERT EINSTEIN'S SISTER), 1924

EINSTEIN'S SKULL IS CLEARLY, AND TO AN
EXTRAORDINARY DEGREE, BRACHYCEPHALIC, GREAT
IN BREADTH AND RECEDING TOWARDS THE NAPE OF
THE NECK WITHOUT EXCEEDING THE VERTICAL.
ASTRONOMER CHARLES NORDMANN, CIRCA 1922

LIFE IN A CIDER BOX

Albert Einstein (1879–1955) is recognized as one of the greatest physicists of all time. What might we learn if we could study the folds of his brain? Although we don't have the brains of such geniuses as Newton or Galileo, scientists today should be able to determine if Einstein's brain differs from the brains of mere mortals. For years, however, his brain was missing. Just thinking about it gives me the same shivers I felt while watching *Night of the Living Dead*.

My latest information indicates that Einstein's brain is in the custody of Thomas S. Harvey, M.D., the pathologist who performed Einstein's autopsy in 1955. For many years, Harvey kept the brain in a cardboard box behind a beer cooler.

It's not clear why Einstein's brain is still around today because the rest of his body was cremated. Some say that his brain was saved at his son's request. Others sources suggest that Einstein himself wanted his brain saved, but Einstein's estate denies this. Contrary to what you might think, research on his brain did not take place until 30 years after his death— perhaps because no one could agree how to cut up this priceless object. In 1978, *Science* magazine suggested that the delay was never satisfactorily explained.

The search for Einstein's brain began in 1976 when *New Jersey Monthly* assigned one of its reporters, Steven Levy, the task of finding the brain. Levy's detective work led him to Thomas Harvey's office in Wichita, Kansas, where Einstein's brain floated in a mason jar packed in a cardboard box marked "Costa Cider." Levy reports that Harvey had little interest because "he had other things to do." In the August 1978 *New Jersey Monthly*, Steven Levy tells his own feelings as he gazed at the chunks of Einstein floating in Harvey's jar:

> I had risen up to look into the jar, but now I was sunk in my chair, speechless. My eyes were fixed upon that jar as I tried to comprehend that these pieces of gunk bobbing up and down had caused a revolution in physics and quite possibly changed the course of civilization. *There it was!*

In 1981, *Science* magazine again asked Dr. Harvey if he had finally finished his research on Einstein's brain. Harvey's response was, "No concrete plan. I have my ideas about it but they have not solidified." *Science* magazine wrote, "Harvey possesses small fragments of the brain but declines to say exactly where they are now stored. Einstein's estate, he says, has no interest in them."

Finally, in 1985, neuroanatomists Marian Diamond and Arnold Scheibel persuaded Dr. Harvey to give them some tissue samples. Having already conducted experiments demonstrating that rat brains are larger when rats have an intellectually stimulating environment, Diamond was curious to see if there was a similar effect on Einstein. After counting cells in Einstein's brain, they found more glial cells than would be found in normal brains. (Glial cells support the neurons, which are the functional units of the brain.) Figure 27 shows "area 9" and "area 39," the two regions of Einstein's superior prefrontal and inferior parietal lobe from which cells were counted in both hemispheres. Diamond and Scheibel chose these areas because they were concerned with "higher" neural functions such as the capacity for abstraction, calculation, planning, and establishment of behavioral strategies, attention, and imagery. Neurons in these areas don't receive primary sensory information but rather interpret impulses sent to them from other areas of the brain. The researchers suggest that Einstein

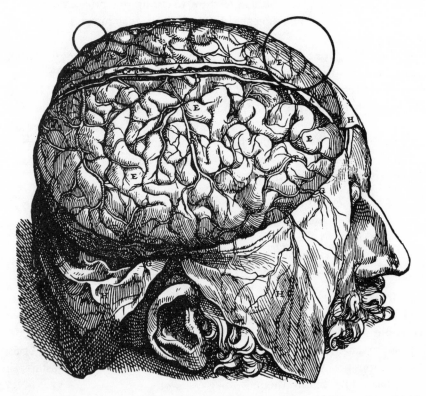

27. A schematic representation of Einstein's brain indicating regions from which samples were removed for cell counts.

had more glial support cells because his neurons (the workhorses of the brain) had greater metabolic need as a result of his unusual conceptual powers.

Oddly enough, Diamond and Scheibel found Einstein to have *fewer* glial cells than expected in area 39 of Einstein's left brain hemisphere. Could Einstein's low number of glial cells account for his slow speech development? We know that this area impacts speech because lesions in this area lead to dyslexia. Interestingly, Einstein once mentioned that written and spoken words were not important when he formulated his theories. Imagery and emotion actually played a greater role. In addition, Einstein had trouble learning to speak when he was a child. He reminisced in 1954:

> My parents were worried because I started to talk compara-
> tively late, and they consulted the doctor because of it. I cannot
> tell how old I was at that time, but certainly not younger than
> three.

Even when Einstein was nine years old, he was still behind in speech development, and later in life he acknowledged his own "poor memory of words." Straus, one of Einstein's associates, remarked in 1979:

> Einstein said that when he was between two and three years
> old he formed the ambition to talk in whole sentences. If some-
> body asked him a question and he had to answer, he would
> form a sentence in his mind and then try it out on himself,
> thinking that he was whispering it to himself. But, as you know,
> a child is not very good at whispering so he said it softly. Then
> if it sounded all right, he would say it again to the person who
> had questioned him. Therefore, he sounded, at least to his
> nursemaid, as if said everything twice, once softly and once
> loudly, and she called him "Der Depperte," which is Bavarian
> for "the dopey one."

Today, most of Einstein's brain is sectioned for microscopic study. My sources tell me that only his cerebellum and a piece of his cerebral cortex remain intact, drifting in a formaldehyde-filled jar under a beer cooler in Dr. Harvey's office. If Einstein hoped that his whole brain would be preserved for future revival by advanced technology, he would likely be disappointed.

WHERE ARE EINSTEIN'S CHILDREN?

Have you ever wondered if Albert Einstein had children, and if so, what became of them? I was able to track down specific information regarding one of his sons from a book published in 1991. Hans Albert Einstein (1904–1973) was the first of Albert Einstein's two sons. He was educated in Zurich, became a hydraulic engineer, and joined the faculty of the University of California at Berkeley in 1947. In 1959 he married his second wife, neurochemist Elizabeth Roboz, who wrote *Hans Albert Einstein: Reminiscences of His Life and Our Life Together* (see Further Reading).

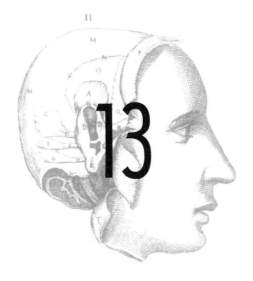

DO WE REALLY USE
ONLY 10 PERCENT OF OUR BRAIN?

SCIENTISTS WHO TELL US WE USE ONLY 10
PERCENT OF OUR BRAINS, ONLY USE 10 PERCENT
OF THEIR BRAINS.

ROLLIN DENNISTON, PSYCHOLOGIST

OUR BRAINS HAVE EVOLVED TO GET US OUT OF THE
RAIN, FIND WHERE THE BERRIES ARE, AND KEEP US
FROM GETTING KILLED. OUR BRAINS DID NOT
EVOLVE TO HELP US GRASP REALLY LARGE
NUMBERS OR TO LOOK AT THINGS IN A HUNDRED
THOUSAND DIMENSIONS.

RONALD GRAHAM, DIRECTOR OF INFORMATION
SCIENCES RESEARCH AT AT&T RESEARCH

IN THE BRAIN, MEMORIES LIE OPENED, ONE INTO
THE OTHER: THE CRUNCH OF THE AX AS HE SWINGS
DOWN HARD, HIS WIFE CALLING HIM IN, A WOMAN
SINGING HIS NAME IN THE DISTANCE. . . . BY NOW
THE WIFE MAY BE DEAD TOO, THE AX PASSED DOWN
TO HIS SON, OR RUSTED UNDER THE WOODPILE.
THE WOMAN CANNOT RECALL HER OWN CLEAR VOICE
OR THE FEATURES OF THE MAN WHO SHOULD BE
BONES.

SUSAN BARTELS LUDVIGSON
ON LEARNING THAT CERTAIN PEAT BOGS CONTAIN
PERFECTLY PRESERVED BODIES

THE 10 PERCENT MYTH

There's an old adage that we use only 10 percent of our brains. Where does this come from? Does it mean that 90 percent of our neurons never fire or that we have some vast untapped mental resource?

Some have attributed the 10 percent figure to psychologist and philosopher William James (1842–1910), who said, "We are making use of only a small part of our possible mental and physical resources." Anthropologist Margaret Mead actually used a figure of six percent when referring to the brain's capacity.

The "10 percent myth" was popular in the 1920s. For example, the 1929 World Almanac ran an advertisement from a self-improvement company that stated, "Scientists and psychologists tell us we only use about 10 percent of our brain power." Despite claims like this, most brain researchers today say that there's no evidence that we have immense untapped resources of brain power in our 10 billion neurons.

In the 1920s, scientists *understood* the function of only about 10 percent of the brain. For example, the "brain" entry in the 1911 edition of the *Encyclopaedia Britannica* stated, "It is true that the greater part of the cortex remains still terra incognita unless we are content with mere descriptive features concerning its coarse anatomy." The 10 percent myth may have arisen from the fact that 90 percent of the brain's function was unknown in earlier times.

Additional promotion of the 10 percent myth has come from individuals interested in parapsychology, for example, people intrigued by ESP, and psychokinesis. They suggest enlightened people tap these resources to accomplish great feats. Barry Beyerestine from Simon Fraser University's Brain Behavior Laboratory wrote in a 1988 *Skeptical Inquirer*:

> Origins of the 10-percent myth are obscure, but the concept was widely disseminated in courses like Dale Carnegie's and canonized in public utterances by no less a personage than Albert Einstein. I believe the error arose from misinterpretations of research in the 1930s showing that, with evolutionary advancement, a progressively smaller proportion of the brain is tied to strictly sensory or motor duties. For methodological reasons, the enlarged non-sensory, non-motor areas were referred to as the 'silent cortex,' though they are anything but silent. They are responsible for our most human characteristics, including language and abstract thought. Areas of maximal activity shift in the brain as we engage in different tasks, and there can be some reorganization of functional regions after brain damage; but there are normally no dormant regions awaiting new assignments.

A response came from James W. Kalat of North Carolina State University's Department of Psychology a few months later in the *Skeptical Inquirer*:

> That's a plausible explanation, but I have found another possibility that I like better. Consider the following quotation from R.S. Woodworth, *Psychology*, 3rd ed., Henry Holt 1934, p. 194: "The total number of nerve cells in the cerebral cortex is estimated to be about 14,000,000,000. Many of these are small and apparently undeveloped, as if they constituted a reserve stock not yet utilized in the individual's cerebral activity." The quotation suggests that all the brain's small neurons (whose function was then unknown) might simply be "baby" cells waiting for a chance to grow and then start contributing to cerebral activity.

Finally, the 10 percent figure may refer to the percentage of our brain responsive to our conscious command. A large part of the brain consists of suborgans such as the limbic system, cerebellum, and medulla, which run our body autonomously and subconsciously. If we were to somehow temporarily access this subconscious system for a bit of erudite philosophizing, we would die! Despite what transcendental meditators might tell us, 90 percent of the brain is actively engaged in keeping us alive and is not directly accessible to cortical control.

Columnist and author Cecil Adams suggests that we must have *some* spare brain cells or else no one would recover from strokes. After age 30, we may be losing around 100,000 brain cells a day. Between early adulthood and age 90, our cortex loses around 20 percent of its neurons, although the remaining 80 percent develop more interconnections possibly in an attempt to combat the deterioration.

At any one given time, only about five percent of the brain's neurons appear to be active, but researchers have not found some unused portion of the brain. Our three-pound brain is just too small, uses too many of the body's resources (such as oxygen), and has too much to do for 90 percent of it to be in a vegetative state.

It is probably true that our brains can be exercised and thus grow thicker and more structurally complex. Rats raised in stimulating environments not only have thicker cerebral cortexes and larger neurons, but also have more connections between neurons and more glial (support) cells. Presumably, humans who live in stimulating environments have brains with more interconnections than those who do not.

"IS YOUR BRAIN REALLY NECESSARY?"

Perhaps some support for the 10 percent myth comes from studies on intelligent children with tiny brains resulting from hydrocephalus, a condition where cerebrospinal fluid expands the brain's internal ventricles and compresses brain tissue. Dr. John Lorber, a Professor of Pediatrics at the University of Sheffield, has performed CAT and EMI scans on over 500 young people with huge ventricles and very thin residual brains. Interestingly, these people with mere shells of a brain have no physical defects and have normal intelligence. Some are actually very intelligent with high IQs. Dr. Lorber suggests that hydrocephalus can slowly progress over many years without ever causing symptoms. Specific functions of the brain, such as the motor cortex, may relocate to other regions starting from early infancy. Another possibility suggested by Lorber is that "we do not need such a large quantity of brain and only need to use a very small part of it under normal circumstances."

In a 1978 volume of *Archives of Disease in Childhood*, Dr. Lorber wrote about brilliant people with no neopallium (the phylogenetically youngest portion of the brain, which appears first among the more advanced reptiles and has become the largest part of the mammalian brain):

So far some 70 individuals between 5 and 18 years of age were found to have gross or extreme hydrocephalus with virtually no neopallium who are, nevertheless, intellectually and physically normal, several of whom may be considered brilliant. The most striking example is a young man of 21 with congenital hydrocephalus for which he had no treatment, who gained a university degree in economics and computer studies with first class honors, with an apparent absence of neopallium. There are individuals with IQs of over 130 who in infancy had virtually no brain and some who even in early adult life have very little neopallium.

I personally have seen CAT scans of these people with "potato-chip" brains. Each patient has a thin layer of brain tissue lining his or her skull. Many of the potato-chip people are enrolled in universities and doing very well. What Lorber's remarkable cases demonstrate is the amazing ability of the brain to adjust to massive disruptions, providing they occur slowly enough and early enough in life. Therefore, although scientists do not believe that 90 percent of our brain capacity lies dormant, it is true that the brain has remarkable powers of recovery when substantial brain tissue is lost at an early age of life.

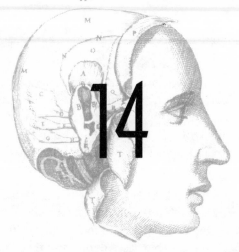

THE HUMAN MIND QUESTIONNAIRE

THE MEASURE OF MAN IS NOT HIS INTELLIGENCE. IT
IS NOT HOW HIGH HE RISES IN THIS FREAKISH
ESTABLISHMENT. THE MEASURE OF MAN IS THIS:
HOW SWIFTLY CAN HE REACT TO ANOTHER PERSON'S
NEED. AND HOW MUCH OF HIMSELF CAN HE GIVE?

PHILIP K. DICK

MAYBE THE BRILLIANCE OF THE BRILLIANT CAN BE
UNDERSTOOD ONLY BY THE NEARLY BRILLIANT.

ANTHONY SMITH
THE MIND

Throughout this book I've been using the terms "genius" and "intelligence" without much discussion or definition. In this chapter we digress and informally examine various facets of the human mind relating to intelligence.

PLANET DUMB AND PLANET SMART

Imagine a planet called DUMB where there are never any geniuses, where IQ is limited to 100. Planet DUMB is identical to Earth in all respects except for this limitation on human intelligence. What would life be like on DUMB? How would civilization be changed if human IQ had been clamped at 100 since the dawn of humanity? On the other hand, what would civilization be like on planet SMART where human IQ had always been 40 points higher? These are just a few of the questions I asked psychologists and other scientists interested in the effects of intelligence on civilization. My questions were asked through computer mail and often posted to computer bulletin boards. One common source for such information exchange was "sci.psychology," an electronic bulletin board or "newsgroup" that is part of a large, worldwide network of interconnected computers called Usenet. The computers exchange news articles with each other on a voluntary basis.

The relationship between human intelligence and the evolution of human societies fascinates both laypeople and seasoned psychologists. The eight questions I asked were as follows:

1. What is genius?
2. Marilyn vos Savant is listed in the Guinness Book of World Records as having the highest IQ in the world—an awe-inspiring 228. How would humanity be affected by the presence of an individual with an IQ ten times greater than hers?
3. What new areas of human thought might be open to this hyperIQ person? What profound concepts or areas of awareness might be available to which we are now totally closed?
4. When Marilyn vos Savant was asked how limited or unlimited she thought brain function is, she responded in her book *Ask Marilyn*: "I suspect it is far more limited than we would like to believe. A fish is unaware that it can't understand spoken language like a dog, a dog is unaware that it can't understand mathematics like a human, and a human is unaware that it can't understand . . . who knows what?"

    My question is: If a human developed a mind as superior to an average mind as ours is to a chimpanzee's, what effects would this have on humanity?

(Note that we can hardly imagine a chimpanzee understanding the significance of prime numbers, yet the chimpanzee's genetic makeup differs from ours by only a few percentage points. These minuscule genetic differences in turn produce differences in our brains. Additional alterations of our brains would admit a variety of profound concepts to which we are now totally closed.)

5. What would the world be like today if there were never geniuses? What would our world be like if humans had never developed an IQ above 100? What would our world be like if all human IQs were 40 points higher since the dawn of humanity?
6. The brain is obviously a finite organ. How many memories can we accumulate? What percentage of our memory capacity do we use?
7. Would you pay $5000 if it would increase your IQ 100 points? 1000 points?

I present the results at the end of this chapter. As you ponder these questions, think about what the world might be like if geniuses such as those in this book formed a higher percentage of the population. Would this be positive, or would we have too many like Geoffrey Pyke practically going mad with ideas? Recall how, for Pyke, new idea bred new idea, and each new plan flared up so quickly that he could not capture it on paper before the next scheme flickered in his mind. For Pyke, little seemed to get finished because there was always a new thought to chase down, a newer plan to develop because it seemed more important than the one that had preceded it.

WHAT IS IQ?

"WHAT IS A HUMAN BEING, THEN?"
"A SEED."
"A . . . SEED?"
"AN ACORN THAT IS UNAFRAID TO DESTROY ITSELF
IN GROWING INTO A TREE."

DAVID ZINDELL
THE BROKEN GOD

My interest in human intelligence began as a teenager. I remember going into my father's study and gazing at his collection of science fiction books. There was one short story with which I became particularly enthralled: Howard Fast's "The First Men." In the story, scientists attempt to create a new race of humans by obtaining genius infants from all over the world and raising them together in an enlightening, stimulating environment. (Imagine Tesla, Heaviside, Galton, and Bentham raised together from infancy by the most caring, stimulating, and supportive parents.) As the children mature into adolescents, they become very different from ordinary humans. In the end, the geniuses create an impenetrable wall around their little town because they (correctly) fear that ordinary people want to destroy them.

Before addressing the various questions in the Human Mind Questionnaire, I would first like to give some background. For example, when conducting my informal survey, many of the respondents questioned my use of the term "IQ" to indicate intelligence, and I replied that I was using the term loosely to indicate the degree to which humans can make abstractions, learn, deal with novel situations, and understand mathematics. In other words, when I use the term "high IQ," I generally mean high intelligence involving the ability to reason, solve problems, plan, and comprehend complex ideas. In classic IQ tests, the IQ value is obtained by dividing an individual's mental age (determined by tested skills) by his chronological age and multiplying by 100. In particular, an individual's IQ is determined by comparing his test results with an average score achieved by a group of individuals in the same age group. As a result, about the same number of individuals have scores lower than 100 as above it. A score above 130 is considered gifted. The term "genius" refers to those with IQs greater than 140. A score below 80 is considered mentally deficient or retarded. IQ test scores rise until chronological age reaches the middle or late 20s, then scores gradually decline.

Critics have argued that IQ tests favor groups from more affluent backgrounds, and consequently other intelligence tests have been considered. Some have even tried to measure IQ directly using electroencephalographs of the brain's electrical activity, but this approach is still experimental. Some researchers, such as Howard Gardner at Harvard University, suggest that there are really seven broad forms of intelligence: linguistic, spatial, logical-mathematical, musical, bodily-kinesthetic, intrapersonal, and interpersonal. Obviously, standard IQ tests do not test for "intelligence" in all these areas.

Children's IQs are correlated with the social status of their father's occupations. The average IQ of children born to fathers with professional jobs is 115. Those born to fathers with unskilled labor jobs average a 94 IQ. Mean IQs for a range of jobs are listed below:

1. Semiprofessional and managerial, 112
2. Semiskilled, minor clerical and minor business, 104
3. Farm owners, 98
4. Slightly skilled workers, 97

(Note, however, that "lower-class" boys often achieve higher scores in mechanical ability than do "upper-class" boys.)

For various reasons, IQ scores should be interpreted carefully and on an individual basis. An IQ for a person who has grown up in a ghetto with poor schooling does not have the same meaning as the same IQ score for someone who has grown up in an upper-middle-class suburban environment with well-endowed schools. An IQ for someone whose first language is not English but who takes the test in English does not have the same meaning as for someone whose first language is English. Also, "test-anxious" individuals may not perform well on IQ tests.

Given all this, low IQ is still one of the strongest characteristics of criminal offenders as a whole (see Further Reading at end of book). The average IQ of convicted lawbreakers is 92, some 8 points below the population average. (Offenders who "get away" with crime have about the same IQ as those caught.) IQ scores drop sharply for serious, repeat offenders. Individuals raised by abusive, unstable parents or living in crime-infested neighborhoods but who refrain from delinquency often have above-average IQs. Poor verbal ability seems to be the "active ingredient" for crime in the overall IQ score. Boys with low IQ scores frequently act aggressively and impulsively starting at the age of three. Low scores may reflect an inability for some children to express themselves and remember information.

Although IQ is often correlated with many personal achievements, including success at work and school, law-abidingness, income, and social competence, there is sometimes a downside to high IQ (see Further Reading). People with high IQs may have difficulty in forming satisfying relationships. High IQ women sometimes feel anxious, dwell on their thoughts, and often experience guilt. Some men with high IQs are very critical of others, uneasy with sexuality, and emotionally dry.

Interestingly, individuals tend to perform equally well or poorly in the verbal and spatial parts of today's IQ tests. Perhaps this is a new development of the mind. People of the most recent generations have had the freedom to choose their own marriage partners, and this means that married partners are more likely to be alike in intelligence than couples from a century ago. In industrialized nations, couples often meet at colleges, and those who do not go to college meet each other in hometown neighborhoods. Psychologist David Buss from the University of Michigan believes that after a few generations this selection activity produces a population that shows progressively larger individual differences on IQ tests. (Francis Galton, discussed in Chapter 7, would have liked this theory.) If a psychiatrist were transported back to the days of the Cro-Magnon man, he would find much less similarity between the scores achieved by individuals on various IQ scales.

Finally, for those of you interested in the "transhumanist" school of thought, which attempts to predict the consequences to creating systems more intelligent than contemporary humans, see the World Wide Web site: http://www.nada.kth.se/~nv91-asa/trans.html. Transhumanism is a philosophy suggesting that humanity can, and should, strive to higher physical, mental, and social levels. The philosophy encourages research into such areas as life extension, cryonics, nanotechnology, physical and mental enhancements, uploading human consciousness into computers, and megascale engineering.

HOW MUCH INFORMATION CAN OUR BRAIN HOLD?

One of my last questions in the Human Mind Questionnaire deals with memory. I asked this memory question because our storage and retrieval abilities are as astonishing as our intelligence. Some experts have estimated that we can store 10 times more information than is contained in the Library of Congress. No one knows how much brain tissue is involved in memory. Neuropsychologist Karl Lasher destroyed pieces of rats' brains in hopes of localizing memory. After he trained rats to do particular tasks, he removed specific chunks of their brain, put them in the position to perform their tasks, and found the memories always persisted.

The human brain weighs about three pounds and is made of roughly 10 billion neurons, each neuron receiving connections from perhaps 100 other neurons and connecting to still 100 more. The web of interconnections is so complex that the whole cortex can be thought of as one entity of integrated activity. Many neurobiologists believe that memory, learning, emotions, creativity, imagination—all the unique elements of human character—will ultimately be shown to reside in the precise patterns of synaptic interconnections in the human brain. The importance of the brain's system of pathways has led some scientists to hypothesize an equation for consciousness itself: $C = f_1(n)f_2(s)$. Consciousness C is represented on the cellular level by a function of neural cell number n and connectivity s.

Small systems of under 10,000 neurons, such as those in simple invertebrates, are capable of learning and memory. Information theorist John von Neumann once estimated that the memories stored in the human brain during an average lifetime would amount to approximately 280 quintillion (2.8×10^{20}) bits, assuming nothing is ever forgotten.

There are many examples of individuals who display remarkable memories. In 1974, one individual recited 16,000 pages of Buddhist texts without error. In 1987, Hideaki Tomoyori recited the digits for pi from memory to 40,000 places in 17 hours. (Note that in 1987 a NEC SX-2 supercomputer *calculated* pi to more than 134 million digits. In 1989, the Chudnovsky brothers, two Columbia University mathematicians, computed over one billion digits of pi using a Cray 2 and an IBM 3090-VF computer.)

As an example of computational capabilities of the human brain, Willem Klein in 1981 was able to extract the 13th root of a 100-digit number in about one minute. Autistic savants (people who can perform mental feats at a level far beyond the capacity of a normal person but whose overall IQ is very low) are sometimes capable of prodigious feats of memory and calculation. For example, in 1844, Johann Martin Zacharias Dase (1824–1861), a human computer, calculated pi correctly to 200 places in less than two months: 3.14159 26535 89793 23846 26433 83279 50288 41971 69399 37510 58209 74944 59230 78164 06286 20899 86280 34825 34211 70679 82148 08651 32832 06647 09384 46095 50582 23172 53594 08128 48111 74502 84102 70193 85211 05559 64462 29489 54930 38196.

Dase had an incredible brain. He could give the number of sheep in a flock after a single glance. He could multiply two 8-digit numbers in his head in 54 seconds, two 40-digit numbers in 40 minutes, and two 100-digit numbers in 8 hours! Dase performed such computations for weeks on end, running as an unattended supercomputer. He would break off his calculation at bedtime, store everything in memory, and resume calculation after breakfast.

BRAIN SIZE AND INTELLIGENCE

Human intelligence is not correlated well with brain size. In fact the brain of great mathematician K. F. Gauss proved to be an embarrassment to those who thought brain size indicated intelligence. Gauss' brain weighed 1492 grams, only slightly more than average. However, Gauss' brain is more richly convoluted than average brains. Stephen J. Gould's book *The Mismeasure of Man* gives many examples of large-brained criminals and small-brained men of imminence. The largest female brain ever weighed (1565 grams) belonged to a woman who killed her husband, and autistic people have heavier brains than average. (Intelligent people with ultrathin "potato-chip" brains were discussed in Chapter 13.)

HUMAN MIND QUESTIONNAIRE: THE RESULTS

With this short background to intelligence and memory, I now present the answers of psychologists and other scientists to the eight questions in the Human Mind Questionnaire. For brevity, I use the term "hyperIQ" person to denote someone who has ten times the IQ of Marilyn vos Savant (see question 2). A "superIQ" person (question 4) has a mind as superior to an average human mind as a human mind is to a chimpanzee's.

Larissa S. from the Department of Psychology at the University of California, Riverside, is an expert in psychometrics, personality disorders, and forensic psychology. She comments:

1. *HyperIQ person's effect on humanity*: This person's effect would depend on the individual and his or her motivations. If the person was unmotivated to do anything with that intellect, the effect would be little. If the person was interested in technological or medical innovation, there would probably be a great effect.

2. *Effect of superIQ person on humanity*: Some psychologists believe that mental processes are restricted by the language which shapes them. For instance, many psychological constructs were developed in Germany because there were no words for them in English; the things you can think about are limited by the words and constructs you can use to shape those thoughts. vos Savant was, I suspect, referring indirectly to this hypothesis. So in answer to your question: a mind as superior to the average mind as ours is to a chim-

panzee would still be bound by the same linguistic and cognitive capabilities as our own minds. There would probably come a point of diminishing returns such that unlimited intellectual capacity did not translate into unlimited vistas of thought.

3. *$5000 for 100 IQ points?* I would pay this amount. Of course, if everyone who could afford to did, we would have an even worse social stratification than we do now, since the people with money and the people with intelligence would be pretty much the same.

Stephanie H., a freelance writer and editor, has studied various facets of psychology and mind. She comments:

1. *What is genius?* Genius is the ability to draw relationships between disparate ideas or facts.

2. *HyperIQ person's effect on humanity*: A hyperIQ person would have an adverse effect on humanity. Every great thinker in human history has been ridiculed to a certain degree by "average" thinkers. Eventually, society benefits from high IQ people, if the high-IQ people are disposed to conduct research or invent. However, if a great thinker wants merely to offer his or her philosophical musings, society will rarely take notice or fully understand such ideas. (Almost every great philosopher has been misunderstood.)

3. *HyperIQ person's world-view*: The most highly technical minds eventually come to profound theories regarding spiritual issues, so a hyperIQ will better understand the motivations or processes of God.

4. *Effect of superIQ person on humanity*: Humans struggle with the "meaning of life" and "why we die" questions because our brains are poised just at the brink of understanding or perceiving other dimensions of reality. Our brains are limited in how they can obtain knowledge. I believe there is a plane of reality wherein knowledge (i.e., "truth") simply exists and is accessible by all without utilizing conscious modes of perception or understanding. The brain is physically limited by gravity, the density of matter, physical proximity, etc.

5. *Changing IQ, impact on civilization*: If the world never had geniuses, we'd be less evolved, but much happier. My mother, a painfully smart woman, always said that intelligence is a burden. However,

geniuses are cosmic fork-prods for humanity. Though we can rarely appreciate genius in the microcosm, we are able in the macrocosm to improve the mean intelligence bit by bit with the help of an occasional genius mind.

6. *Memory capacity of brain*: The brain's capacity is physically limited, but because we use so little of its memory potential, we'd never encroach upon the finite number of possible memories that we can potentially store.

7. *$5000 for 100 IQ points?* I would not pay to increase my IQ. A person would need an entire childhood to learn to harness and cope with such an advanced brain. To acquire intelligence so abruptly as an adult would be overwhelming and mentally and emotionally crippling.

Dan P. is an IBM research scientist with an interest in human intelligence. He holds a doctorate in theoretical physics and has the following to say about the questions I posed:

1. *What is genius?* Genius isn't raw societal success, nor does it correlate strongly with wealth or even ambition. Genius-level IQs can be found in various occupations (I fondly recall a Mensa belly dancer who had an IQ in the 180s). IQ tests can be used to find individuals who solve puzzles fairly well. Is that genius?

2. *HyperIQ person's effect on humanity*: I don't know if a hyperIQ would have an impact on humanity. There are many dumb politicians who have had much more impact than some very smart Mensa folks.

3. *HyperIQ person's world-view*: What might an idiot savant see that we can't? Would a hyperIQ's mental capacity interfere with emotional development and make him unable to participate in our society?

4. *Effect of superIQ person on humanity*: A smart dog may have more sensitivity and perception about the world than a retarded human. On the other hand, I've never gotten my dog to do calculus. While my dog will dance, I have never been able to get her to have a sense of rhythm. The intelligences of a dog and a human are more similar than the intelligences of a retarded dog and a normal dog, or a retarded human and a normal human. By the way, I believe that some forms of addictive behavior may be associated with demon-

strations of genius. Many people in Mensa are not involved with particularly amazing career paths; therefore, genius may not be expressed.

Bets L., a manager at a large computer corporation, comments in general about the questionnaire:

> I have no idea what my IQ is. However, I assume it is above average, but less than Marilyn's. Even with this simple "above average" capability, I find that I am sometimes frustrated at living in a "C+ society." I have a good friend who is at least Mensamaterial, but she was the loneliest person in high school, perhaps suicidal. Perhaps our society is such that it is still worse to be bright and female . . .
>
> While I would not want to decrease my intelligence, I don't know that one's happiness or well-being is increased simply by increased intelligence. I am not sure that society is better off because of Marilyn's intelligence. I think that inventions have more value to society than intelligence per se. I have seen Marilyn's newspaper column, and I am less impressed with its beneficial impact than I am with a simple "invention" like George Washington Carver's peanut butter . . .
>
> This *is* a C+ society, and because of that, intelligence without practical application does not have much impact.

Syed Jan A., author of *Symmetries of Islamic Geometrical Patterns* and Professor of Mathematics at the University of Wales, comments:

1. *What is genius?* A genius has the capacity to escape the traps of the current models and paradigms. High IQ is not genius. A person with a high IQ may or may not be a genius. A genius may or may not have a high IQ. Was Christ a genius? Did he have a high IQ?
2. *HyperIQ person's effect on humanity*: People with high IQ do not behave in some uniform way. I know one high IQ person who has published 400 papers, but he is totally useless in the area of social relations and solving his problems of employment. He has just had his professorship terminated. I know another person with a very high IQ who can absorb knowledge at a phenomenal rate. He is the

kind of guy who will hack into a Pentagon file. However, he has no discipline or commitment. If a hyperIQ existed, we could not predict how this will affect the world.

3. *HyperIQ person's world-view*: A hyperIQ person will see entirely new "patterns" and regularities in one or more areas of human endeavor. Mathematics and all the other hard sciences plus music are strong candidates for quantum leaps. Literature, art, and religion are not. I agree with Marilyn.

4. *Changing IQ, impact on civilization*: If there were never any geniuses then we would have become extinct. If humans never had an IQ above 100, then we would be existing at the basic animal level, slightly removed from the apes. If our IQ had been 40 points higher then we would, by now, have left the Earth and traveled out into space.

5. *Memory capacity of brain*: Not all of our memories are in the brain. The memories may be stored elsewhere, and the brain may just be a read/write head.

6. *$5000 for 100 IQ points?* Yes, yes, yes, please.

George M., a Professor of Mathematics and Computer Science at the University of Maine, comments:

> The emphasis on IQ is misplaced since it is a relative measure rather than an absolute one. It is based on how well you do at your chronological age. Thus you can have two "equally" smart people, considering that they correctly answered the same questions, but one has a higher IQ than the other because he is younger. It is also interesting to ask how anybody can earn a higher IQ on a test than that possessed by the designers of the test. Presumably the designers had enough intelligence to answer the questions.

Joseph P. is a mainframe computer programmer from San Jose, California, with an intense interest in the human mind. He comments:

1. *What is genius?* A genius is a person who can look at the same thing as you and draw conclusions that you could not. Richard Feynman was a "super" genius who drew conclusions that only a genius could follow. I think our IQ 2280 person would be like Feynman in

that respect. I don't think we could tell the difference between Feynman and the hyperIQ person.

2. *HyperIQ person's effect on humanity*: The presence of a single individual of IQ 2280 would probably be publicized as an oddity. If a mating pair existed, they would probably be perceived as a threat and destroyed.

3. *HyperIQ person's world-view*: The hyperIQ would have emotional difficulties. He/she still has to eat and find shelter, and also has to work with tools and routines meant for less intelligent people. This may cause an intense conflict depending upon the hyperIQ's upbringing. The hyperIQ would spend time exploring deeper meanings and relationships between daily occurrences. As for branching off into new areas of thought, what would the hyperIQ use for stimuli to identify the new areas?

4. *Effect of superIQ person on humanity*: The superIQ will search for other superIQs (human or non-human), or hide the difference to fit into society and find companionship. The superIQ might try to improve humanity in small increments.

5. *Changing IQ, impact on civilization*: If there were never geniuses, the cockroaches and ants would rule the world. If humans had never developed an IQ above 100, humanity would be stuck at a Stone Age level. If all human IQs were 40 points higher since the dawn of humanity, there would be little effect on humanity's development. Cold weather kills a 140 IQ person just as easily as a 100 IQ person. Bears and other predators can't tell the difference.

6. *Memory capacity of brain*: If the brain is like holographic memory, that, is multiple copies of each memory residing in different areas of the brain, then without editing of the memories, the human brain would overload by age one year. I expect the best humans can do is learn to edit and discard unneeded memories, and if they can't then they are considered mentally disabled. Humans use 100 percent of their memory capacities with varying amounts of editing to discard "noise" and to selectively enhance desired memories. The success of discarding and enhancing determines how successful the person will be.

7. *$5000 for 100 IQ points?* Pay to increase IQ? I don't think so. IQ tests are arbitrary. The money could be better invested elsewhere. The hyperIQ person would be isolated from "normal" society. Loneliness would probably kill the person.

Jerry K. is a graduate student in theoretical computer science currently working in algorithmic number theory and computational algebra. He is interested in computational complexity, cryptography, coding theory, and the origins of human intelligence. He comments:

1. *What is genius?* A genius is someone whose intellectual capacities are substantially greater than those of the majority of the population, whose ability to think, remember, and problem solve are far better than most peoples'.

2. *HyperIQ person's effect on humanity:* An IQ of 228 is so far from the mean that its statistical meaningfulness is suspect. For sake of argument, let's assume Marilyn's 228 score really does reflect an enormous intelligence. What has her contribution to society been? The test Marilyn took was administered by a cabal who evaluates tests if returned with a cheque for the appropriate amount. In addition, the test-taker has an arbitrarily long time to complete the test using any available non-living resources for help.

3. *HyperIQ person's world-view:* What is this "IQ"? How was it measured? Of what is it supposed to be a predictor? Without some knowledge of what this means, why would there be any a priori reason that this person would shake our world in any fundamental way?

4. *Effect of superIQ person on humanity:* If we understand the superIQ by the same amount as a fish understands human language, then a superIQ won't have any effect on humanity. Here, vos Savant seems to be saying something reasonably sensible. A pleasant change from her piece on Fermat's Last Theorem.

   As an aside, what does it mean to say a chimp doesn't understand prime numbers? Do those plants whose branch spacing corresponds to numbers in the Fibonacci sequence understand Fibonacci numbers?

5. *$5000 for 100 IQ points?* I'd be better off investing the $5000 in mutual funds or gin and tonics than in elevating an indicator whose meaning is already questionable and in a statistically meaningless range.

William S., an attorney at law in Boston, Massachusetts, has an interest in psychology. His IQ is 160. He comments:

The science fiction novel *The Divide* by Robert Charles Wilson discusses the difficulty with which a hyperintelligent man fits into society. In one scene the hyperIQ recalls the time he had sex with a woman for the first time and, afterwards, wonders whether what he'd done constituted bestiality . . .

Would I pay $5000 to increase my IQ 100 points? Yes. The intellectual boost will allow me to make back more than $5000. More importantly, I'd want to be smarter. I've an IQ of about 160, and the more processing power I have, the better. At least I'd be able to force my damnably autonomic nervous system to learn to touch-type. There are risks of having a hyped-up IQ; these include: (a) serious alienation from "normal" humans or (b) becoming aware of unpleasant truths from which my current relative dumbness has shielded me. However, (a) I'm already fairly alienated, and (b) I'm likewise already pretty depressed as a result of my unenhanced mind's observations of reality and conclusions drawn therefrom, so I'm willing to take the risks. IQ enhancement could not make my life noticeably worse, and the potential for making it better is strong.

Walt J. is a computer expert and pilot from Oswego, New York, and he has an interest in the field of intelligence. He comments:

1. *What is genius?* Genius is connectivity: the imagination to take "A," "B," and "C" and see a potential relationship.
2. *HyperIQ person's effect on humanity*: A person with an IQ of 2280 would have almost no impact on the world for he/she would be unable to communicate with us animals. The hyperIQ's thought processes and concepts would be to us as the study of calculus is to our cats and dogs. A single hyperIQ individual would be insane before adulthood. A group of hyperIQs would be executed or jailed as they established communication with each other. Any sufficiently advanced science/thought is indistinguishable from magic, and fear of the unknown is universal.
3. *HyperIQ person's world-view*: Insights into the world of mathematics would lead to new areas of awareness with resulting impact on other disciplines. We can only guess at the impact in the same manner that my kitty cat can guess at the next step in the computer revolution.
4. *Effect of superIQ person on humanity*: Marilyn states this well. A gradual evolution in intelligence will yield the best progress. Mari-

lyn has obviously been able to communicate at some level with the great unwashed. An intelligence able to communicate with today's hyperintelligent would be able to have the most impact. A hyperIQ could develop ideas on which other intelligent beings could work. The hyperIQ would be our mentor.

5. *Changing IQ, impact on civilization*: If there were never any geniuses, there would be no creativity, and we would be like beasts. If our IQs had been limited to 100, we would revert to the early days of the industrial revolution. The pace would be slower. Creativity is somewhat constrained by politics, and most of the technical progress is within the last 100 years. If we were all elevated by 40 points, every problem you can possibly think of would be solved, but not much else. We'd still be constrained by politics.

6. *Memory capacity of brain*: Short term memory is clearly different from long term memory. We use 90–100 percent of our short term memory, and as we age the short term capacity decreases. Long term memory varies to such an extent between people, that I suspect we use significantly less than 100 percent of it.

7. *$5000 for 100 IQ points?* I would pay for an increase of 100 points without any hesitation. Is Marilyn single? I would hesitate to pay for 1000 points. Let's negotiate for 150–200 points. I would fear the loss of communication with the rest of the world, and I love my kitty cats. The ability to make a significant contribution to the world is tempting, but I need to be able to communicate with the world.

Larry K. is an assistant to the U.S. President, for Professional Development. He comments:

1. *What is genius?* See my answer to the last question.

2. *HyperIQ person's effect on humanity*: A hyperIQ would either be tremendously unhappy, tremendously unpopular, or both. Such a person would die young, and his or her contributions wouldn't be appreciated for hundreds of years, if ever. Such is human nature.

3. *HyperIQ person's world-view*: In Isaiah 55:6–9 we find: "Seek ye the Lord while He may be found, Call ye upon Him while He is near; Let the wicked forsake his way, And the man of iniquity his thoughts; And let him return unto the Lord, and He will have compassion upon him, And to Our God, for He will abundantly

pardon. *For My thoughts are not your thoughts, Neither are your ways My ways,* saith the Lord." Similarly, we gain insight into this question from Proverbs 1:7: "The fear of the Lord is the beginning of knowledge; But the foolish despise wisdom and discipline."

4. *Effect of SuperIQ person on humanity:* Could mankind, by deliberate action, attain this superiority? For the answer, see Genesis 11:1–9, especially verses 6 and 7: "And the Lord said: 'Behold, they are one people, and they have all one language; and this is what they begin to do; and now nothing will be withholden from them, which they purpose to do. Come, let us go down and there confound their language, that they may not understand one another's speech.' " If a human were born with a superIQ mind, "he'd be crucified, and people would argue about him for centuries, and a few, by following his teachings, would attain true knowledge."

5. *$5000 for 100 IQ points?* I would not pay for increased IQ. However, I would pay for wisdom and good judgment. Is "intelligence," as measured by IQ, the same thing as: knowledge, common sense, good judgment, wisdom, creativity?

Kim W. from the University of Wisconsin has an interest in Japanese culture, zoology, and the human mind. She comments:

1. *What is genius?* Genius is a combination of intelligence and the will to use it.

2. *HyperIQ person's effect on humanity:* If, like Marilyn, all the hyperIQ did was write a newspaper column, I can't imagine much impact. I can't imagine an individual with a validly-measured IQ of 2280 produced by current society.

3. *HyperIQ person's world-view:* A mere slightly-above-average human like myself can not imagine a hyperIQ's world view.

4. *Effect of SuperIQ person on humanity:* We could not predict the effect on humanity. Many high-IQ individuals I know aren't very smart when it comes to day-to-day functioning.

5. *Changing IQ, impact on civilization:* IQ is an artificial scale. This is irrelevant.

6. *$5000 for 100 IQ points?* I would not pay someone to mess with my head (anymore than I already do).

Martha S., a software developer from Endicott, New York, has also done graduate work in psychology. She comments:

IQ is nothing more than a measure of how a person scores on an IQ test. Genius, as defined by an IQ measurement, has little meaning. A genius is a person who achieves in areas of thought, activity, invention, or writing. A genius IQ might define a probable potential, but until that potential is used to achieve, there is nothing real to measure. IQ is known to be a false measure if an IQ test in English is given to a person who is not fluent in English. Nonverbal IQs are, for the most part, culture oriented. The concept of competition or performance is not strong in some cultures.

Anyone with a certain "minimum" score can achieve significant intellectual success. Children's ability to learn, to do well on tests, and to achieve in school are based strongly on the expectations placed on them when they are very young, perhaps as young as 2–3 months.

1. *What is genius?* Genius is not the existence of potential. It is using that potential to achieve some goal. A dog who survives in a big city, filled with major highways, is displaying a genius for surviving. In our world, we most often recognize either artistic (nontechnical) or scientific genius. What we are recognizing is the ability of the so-called genius to perform at something we appreciate but cannot do ourselves. Just expounding ideas does not make a genius. Showing ways in which those ideas can be used, creating ways for testing those ideas, writing those ideas down, and having those ideas serve as "jumping off points" for others—this is what we recognize as genius. Geniuses are unencumbered by others' viewpoints and have the ability to create new viewpoints. Geniuses strive to communicate those new viewpoints and put them to work.

2. *HyperIQ person's effect on humanity:* A person with an IQ of over 200 could be bored by some of the teaching methods and expectations of our world. A person with an IQ over 500 or 1000 or 2000 would be limited by the environment in which they are raised. They would be exposed to our feeble concepts of the physical world, science, math, history, art, philosophy, psychology, music, and literature. I strongly doubt that anyone raised in our culture would be able to conceive from a viewpoint outside of it. Therefore, while hyperIQs might make significant contributions to our culture, and help it move forward more quickly, they would not reverse cultural directions.

The hyperIQ might develop new concepts in philosophy, psychology, medicine, art, music, science, math, or literature. However, such people might also end up in an insane asylum or in a school for mental deficients, since they would not fit the "norm." A hyperIQ child's ideas might well be seen as retarded or insane. A hyperIQ, however, might be intelligent enough to be able to "hide," to appear the same as "average" people.

A hyperIQ might learn to earn a living with the least possible expenditure of energy (for example, writing romance novels), and then use their remaining time to explore areas of personal interest. They may not have an interest in benefiting humanity or furthering of human knowledge. A hyperIQ would be curious about almost everything and take bits and pieces from many different, seemingly unrelated areas and combine them in new, useful ways.

Since we can't comprehend that we are limited by that which we can not conceive, I believe that hyperIQs would still be limited in the same way, since they would be raised by us. The effect of such a mind on the humanity would be negatively correlated with the amount by which its thought patterns differed from ours. If such a mind were to develop a new medicine that could cure cancer or AIDS or prevent Down's syndrome, then the effect of that discovery would have an effect on us.

3. *Changing IQ, impact on civilization*: If there had never been geniuses, the world would not be very different. Mankind itself evolves and creates. If humans had never developed an IQ above 100, then, in order to survive, they may have had to strive more strongly to use what they had. If the average IQ were 40 points higher, there would also be little effect. Most of human history is the battle of survival. This would not have been different.

4. *Memory capacity of brain*: While the brain is physically a finite organ, no one has ever demonstrated the limits to what it can store. Nor has it been demonstrated that there are limits to the number of cross-connections it can make. I believe that there are no practical limits to the number of memories we can store. (If someone ever approaches the physical limit, I believe the brain will automatically develop some form of data compression mechanism. Maybe it already has one.)

5. *$5000 for 100 IQ points?* I would not pay $5000 to increase my IQ. A 100 point gain might frustrate me. I still couldn't make the world into what I want it to be. A gain of 1000 points would be meaning-

less. What would it serve me? Would I ever again be a part of the world? Would I be able to "get by" with less work? Would I belong to this world at all? I don't want to be somebody else, although I would like to be more secure and do things more easily. In this case, a little would go a long way, but I would not pay for an IQ increase, unless I were guaranteed that that increase would benefit me in defined and specific ways.

Clint S. is a Professor of Physics at the University of Wisconsin and author of *Strange Attractors: Creating Patterns in Chaos*. He comments:

Marilyn vos Savant's staff once called me to verify her explanation of Newton's third law, thus demonstrating that the existence of a superior intelligence does not obviate the need for others with lesser abilities.

1. *HyperIQ person's effect on humanity*: Mankind would not be affected by the existence of a single individual with an IQ in the thousands. As much as I admire Marilyn vos Savant, I'm not aware of any significant impact she has had on humanity. More would be gained by a much smaller increase in the IQ of everyone. However, if we had a fixed number of IQ points to distribute, there are individuals for whom a substantial allotment would reap great benefit, while it would be totally wasted or even deleterious for others. We don't, for example, need smarter criminals. I think intellectual advances are not so much limited by our finite IQ but rather by other factors such as ambition, perseverance, and opportunity.

   There is a sense in which endowing a small number of individuals with really superior IQ could be detrimental to society. These people would become oracles, and many of the rest of us would acquiesce to their superior judgment and wisdom. This would further exacerbate an already serious problem, especially in the sciences, of a knowledge polarization between the haves and the have-nots. Vast segments of society have written off science as too difficult or obscure to warrant their study and understanding, and those with the knowledge, especially in the medical sciences, have become almost like gods. The fact that IQs are spread over a relatively narrow range gives everyone the possibility and incentive to contribute something beneficial to society.

2. *Effect of superIQ person on humanity*: Superior intellect dispensed to a small number of people would probably be wasted. In an extreme case, these individuals might become reclusive, indulging them-

selves in reading, calculating, computing, etc., while having zero impact on society. At the least, there is likely to be an impedance mismatch as the results of their activities are translated for the rest of humanity. There have been mathematical geniuses in the past whose work cannot now be understood or replicated.

3. *Changing IQ, impact on civilization*: Evolution is driving us toward increasing intellectual abilities. If IQs could be artificially enhanced, this would accelerate that process. Thus the question of how the world would be different if IQs were higher is equivalent to the question of what the world will be like sometime in the future. Only the rate of progress would be effected. An instantaneous change in the average IQ would only gradually show its effect. If IQs were lower, society would now be something like it was at an appropriate time in the past.

4. *$5000 for 100 IQ points?* Despite my generally skeptical view of the benefits of superior intelligence, I would consider it money well spent if I could raise my IQ by 1000 (or even 100) points for a mere $5000. I would expect to recover the investment easily within the first year. Sign me up.

D. J. is a computer programmer from Raleigh, North Carolina, with an interest in the life of Albert Einstein and Richard Feynman. He comments:

1. *What is genius?* A genius is someone who is really good at something (like physics, art, or making hamburgers) to a degree that amazes the rest of us. Usually this comes from dedication or extreme fascination with the endeavor. Frequently what makes a genius is just the ability to think differently about an old problem, and thereby find a unique solution. That's what distinguished Einstein and Feynman, who were hailed as geniuses in their own lifetimes, even though both had trouble with mathematics in school.

2. *HyperIQ person's effect on humanity*: The IQ is a score on a standardized test. Since we don't know what properties of intelligence this test measures, it is difficult to say what impact a higher score would have on the individual. The range 70–228 describes the difference between a person not able to function on their own in society and Ms. Savant. Someone ten times as intelligent would be

unimaginably smart. However, unless the individual was inclined to unbalance world affairs, the impact of such a person would be negligible.

3. *HyperIQ person's world-view*: A hyperIQ may be able to organize a larger collection of ideas than we can. He could "think through" more complex ideas. This might allow him to reach conclusions more quickly, act in a more certain manner, and make better predictions (from more complex predictive models) about the world around him. It is impossible for us to speculate meaningfully about what concepts or awarenesses are closed to us (since by definition we can't imagine them), so we cannot predict which of them might be opened by increased "intelligence."

4. *Effect of superIQ person on humanity*: Actually, chimpanzees can understand some mathematics, including fractions, but they are not able to generalize their knowledge into strategies for delaying gratification in order to reap a larger future reward. (Perhaps monkeys could be trained to understand prime numbers.) An individual with such superior "intelligence" would only have an impact on the world if he were able to conceal his difference and make his contributions surreptitiously. Overt display of such an alien trait (like a dog displaying advanced calculus to his fellows) would lead us to believe him deranged, and to discount his ravings.

5. *Changing IQ, impact on civilization*: If you define a genius as I have above, it's obvious that human society would have stopped at the first big stumbling block if there were never any geniuses. I disagree with your basic premise that IQ measures intelligence or the ability to think creatively, so the difference between our world and the world of IQ+40 would be that IQ tests would be harder! If people really were more intelligent, we'd each have a shorter schooling period at the beginning of life, or get an earlier, better understanding of the world. Raising an individual's real intelligence may give him or her an advantage in society, but raising everyone's intelligence would only eliminate "Inside Edition."

6. *Memory capacity of brain*: Since nobody knows how memories are laid down, nobody can answer that question. It's entirely possible that we use all our "memory capacity." The brain grows throughout our lives, and it decays throughout our lives. The balance between anabolism and catabolism determines its size and usefulness at any given time.

7. *$5000 for 100 IQ points?* I would pay to have my intelligence increased. But I doubt that that change would be reflected in IQ score.

Curtis K. is a lawyer and expert on the legal problems associated with virtual reality and computer art copyright. He comments:

1. *What is genius?* Douglas Hofstader, in his book *Fluid Concepts*, suggests that the brilliant and creative mental activity that we'd like to model in a computer program is marked by the ability to make creative *slippages* between concepts and categories—reaching for enlightening metaphors and analogies, and then re-examining the facts presented in the new light. See related notions in T. Kuhn's *Structure of Scientific Revolutions*.

2. *HyperIQ person's effect on humanity*: What does IQ measure, short of the ability to take IQ tests? IQ tests may not be able to meaningfully generate figures like an IQ of 300, let alone 2000. We find limits in many other areas as well (temperature, density, the SAT test, the probable life of this universe, etc.). By the way, see O. S. Card's science-fiction novel *Xenocide* which analyzes obsessive–compulsive disorder's relationship to genius. I wash my hands too often. Lady Macbeth was smart, but not a genius.

   Let's not assume that geniuses are always odd; Feynman wasn't. There are many sorts of genius, and nothing suggests that you can generalize. A chess "genius" has no relationship to a law "genius." Of course, you might limit the notion of genius to areas such as mathematics, music, chess, and the like where stunning (almost idiot-savant) abilities are found, but that's arbitrary. (See the biography of Feynman written by J. Gleick.)

3. *Memory capacity of brain*: The brain physically is a finite organ but that doesn't tell you much about its capacity, any more than the finite rules for Go, the English language, or the number of musical notes tells you much about the capacity of the game, the conversations we can have, or the music available to us. Just as a simple mathematical formula can generate an infinite sequence of graphs, the limited raw material of the human brain may not generate an identifiable limit on memory capacity. Memories in the human brain are not stored like bytes in RAM and thus one cannot fill a location with a memory. Sure, some memories are associated with specific sites; and lesions at some sites can wipe memories; but

much of what we call "memory" is generated as needed and is a function of simpler mental bits that have multiple functions. Combinatorial mathematics only gives us the barest hint of the capabilities here.

4. *$5000 for 100 IQ points?* What's the downside? I wonder what the FDA would say to this . . .

Stewart D. is a computer artist from California and interested in mystical theories of intelligence. He comments:

1. *What is genius?* The whole issue of genius is too subjective. We know no intelligence greater than ourselves. We have no objective reference. What is the nature of intelligence? Is there an interface between intelligence and the "soul"? Between intelligence and matter? Between intelligence and the cosmos? Is visual (artistic) genius different from rational genius? Why does humanity seem to assign a different value to artistic genius versus scientific genius? Is it possible for us to judge intelligence?

2. *HyperIQ person's effect on humanity*: Phenomena like this have occurred: Pick a prophet of the ancient world, any saint, heretic. These hyperIQs have had a clear vision of humanity and the cosmos. Would a hyperIQ be able to make his opinions available to the world? Would he be interested in and able to manipulate current media markets? Buckminster Fuller committed "egocide" after his first economic failures. How does a hyperIQ deal with emotion? How will a hyperIQ deal with opposition? What about Marilyn vos Savant? I know little about her; all she seems to do is to "be" and to write. I suppose I should go read what she has written, but is she doing useful work? Is this the "grace" of her intellect, the acceptance of local action even though she knows global solutions? A hyperIQ will not affect humanity unless that person chooses to.

3. *HyperIQ person's world-view*: Most "mysteries" (acts of God, "fate," etc.) may be understood as complexities like physics, cosmology, the global climate, weather, epidemiology, ecology, economics, and social mechanics. Computers, information networks, and genetic algorithms will eventually gain insight into these problems. A hyperIQ could handle them as well. Do solutions to the Langland's program of mathematics or the grand unification theory of physics have immediate impact on humanity? No, but their application

does. To have impact, the superIQ must be a practitioner as well as a theorist. Will recent events like the bombing of Hiroshima prevent future Einsteins from participating in the application of physics? Yes. Will humanity offer a more acceptable outlet for the intellect?

4. *Effect of superIQ person on humanity*: The impact on humanity depends upon the person's choice to implement the solutions.

5. *Memory capacity of brain*: Memory varies wildly between individuals.

6. *$5000 for 100 IQ points?* I would pay for increased IQ if I had $5000 to blow on something with a dubious return. After my experiences in the early 1970s, I am leery of "smart" drugs. What methods would I be using to increase my IQ? I'm not convinced financial success is linked to intelligence.

Stephen P. is a computer consultant for a university computer center. His interests include psychology, philosophy, comparative religion, mythology, science fiction, and fantasy. He comments:

1. *What is genius?* Geniuses have radically above-average capacity and performance in some area of human endeavor, not necessarily intellectual endeavor. There is also a social viability component to genius. You or I might have the capacity to be the greatest stone spear tip creators the world has ever seen, but this skill is not in demand. Thus, we aren't likely to be exposed to the skills; nor would our achievements, even if we were exposed, be likely to receive much recognition. There are some theorists that posit a number of different "intelligences" which cover a wide range of capabilities other than reasoning capacity. In this respect, Hitler was clearly a very intelligent man when directing people's energy.

2. *HyperIQ person's effect on humanity*: I don't think we have any concept of what an IQ of this magnitude would mean. IQ ("Intelligence Quotient") originally derives from Binet's formula of "mental age" divided by "chronological age." It was generally meant to measure the relative development of school-age children. A person's mental faculties are not constant nor do they change consistently over time.

3. *HyperIQ person's world-view*: You are assuming the following: 1) that the hyperIQ has had the background and experiences allowing him to make use of his talents; 2) that the person has sufficient edu-

cation enabling him to apply the intelligence to the world; 3) that the person has the temperament and stability to make effective use of his abilities; and 4) that those around the person are sufficiently supportive. Such a person might be able to discover comparatively simple underlying patterns to human behavior, world events, economic trends, physical phenomena. . . . This could dramatically improve our abilities to deal with these complex events. Much of the history of ideas involves people who discovered a pattern or connection underlying complex or seemingly unrelated phenomena.

4. *Effect of superIQ person on humanity*: The focus on intelligence as being the determining factor in doing great things is overrated. Conversely, someone of only "average" intelligence with the right kind of encouragement can still make a great person. In rebuttal to the assertion that "mentally challenged" folk don't contribute to the development of society or the human race, I once heard it said that: "They also aren't the ones building nuclear bombs or chemical weapons." High IQ is no necessary defense against delusions or fanaticism. Brilliance doesn't imply ethical or moral superiority.

5. *Changing IQ, impact on civilization*: If civilization's IQ were lower, or there were never geniuses, we would have destroyed ourselves by now. (We might still be fighting wars with swords and crossbows.) If average IQs were higher, we'd be living in utopia while reaching for the stars. By the way, several types of "mental illness" have been traditionally linked with high mental capabilities. There is the old saw of genius and insanity being closely linked. Many highly intelligent children are not only mentally gifted, but are also very sensitive, intuitive, and/or empathic at a level uncommon for children of that age. This kind of awareness, in the face of an insensitive world that has trouble comprehending their capabilities and sensitivities, can produce some very wounded folks.

6. *Memory capacity of brain*: There is some evidence that complete multimedia recordings of every experience since birth are in our brain.

7. *$5000 for 100 IQ points?* I would pay for an IQ increase only if assured that I would also get the opportunities and the motivation to make good use of it.

Will W. graduated from the Massachusetts Institute of Technology and is interested in human behavior, secure computer networks, aircraft navigation systems, video games, electronic photography, cryptography, nanotechnology, and software development. He comments:

1. *What is genius?* Give brain science another two or three centuries, and we might have meaningful answers to questions like this. Brains are very complicated, not at all like piston rings or air filters. It's easy to say what makes a piston ring great or terrible. Not the case with brains. It's hard to study genius. Living, working brains are inaccessible. You have to administer anesthetic (which may alter the brain's behavior), and you have to drill messy holes and tether people to lab instruments. Living, working brains have billions of microscopic parts, all speaking a language of their own, and nobody has a Neural–English dictionary. The compilation of such a dictionary would involve identifying relevant abstractions or levels of abstraction. Gradually improving instrumentation will help us make progress.

2. *HyperIQ person's effect on humanity*: HyperIQs may have no effect. I've never heard of Marilyn vos Savant. People are almost never famous for having extraordinary IQs. People who make a lot happen (e.g., entrepreneurs, generals, Mother Teresa) have not been widely reported to have outlandishly high IQs.

   Numerical metrics of intelligence are garbage. They have no practical meaning, and are mostly used by rather dim administrators to screw up the lives of defenseless schoolchildren. Real intelligence is far too subtle a thing to be measured like a temperature. I'm much more interested in people who make a big difference in the world.

   Being very stupid is certainly an obstacle in life, but intelligence per se is no guarantee of success nor of making a large impact on the world. People often wrongly attribute the atomic bomb to Albert Einstein because of the letter he wrote to FDR. His work contributed to the bomb in only a very remote way. Almost all the bomb work was done by people who might be described as "more practical."

   Extra intelligence beyond a basic functional threshold doesn't buy you a hell of a lot. What would you do with it? What does Marilyn vos Savant do with hers? Does she invent things? Does she do AIDS or cancer research? Does she have unusual insights into economics or politics? Is she negotiating peace in the Middle East or Northern Ireland? (It's not my intent to criticize her for failing to exercise her gift in service of humanity. Whether to do so would be her choice. Rather, it's to question the belief that high intelligence naturally translates into making a huge difference in the world.)

Making a difference in the world and solving important problems takes some intelligence, certainly, but there are many other attributes that are at least as important, like ambition, courage, or compassion. Henry Ford once said something along the lines that he didn't need to be a genius if he could employ the services of others who were. The book *Protector* by Larry Niven addresses many questions on intelligence.

3. *HyperIQ person's world-view*: You're asking the reader to second-guess a hypothetical individual who is supposed to be much smarter than the reader, having already presumed that this individual's insights are beyond the grasp of the average guy. I always wondered how Niven reconciled this problem while attempting to write about the behavior of superintelligent beings. There are the standard "big problems of society" to which I alluded earlier: what do we do about war and crime and poverty, how do we cure diseases; how do we organize society to keep it both free and peaceful; how do we handle technological change better; what's the optimal balance between a free market and a federally subsidized safety net, etc. What are the next big insights in those areas? Again there are standard answers: with the growth of electronic information exchange and computer technology in general, many of these things can be expected to work themselves out in the normal course of events, and require no special insights. But there are some opportunities like the Robert Axelrod work on game theory ("The Evolution of Cooperation") which was briefly extended by Douglas Hofstader.

There's a spiritual thing happening these days: people are excited about angels. Maybe this hyperIQ will tell people how to see and talk to angels. I understand the Catholic Church now holds the opinion that angels have free will, which I believe is a reversal of an earlier position. Other spiritual developments include Redfield's book, *The Celestine Prophecy*, and the book *Care of the Soul*.

The hypothetical genius could be the next James Joyce or Salvador Dali. More likely, any reasonably bright person will go out and find his/her own areas of interest. Nobody knows what effect the hyperIQ will have. They'll see the advantages of steering clear of fame, so we probably won't hear about their work until later.

4. *Effect of superIQ person on humanity*: You may be amused to know that we share about a third of our genetic material with a spinach. When I first heard that, I had to think a few seconds to remember that a spinach was even alive. I could make a pretty believable case that we are nearly as smart as it is possible to get. Why chimps don't understand primes is because they don't understand math (beyond perhaps addition of small integers), and the reason they don't understand math is that they have only rudimentary mental hardware for extracting and manipulating abstractions. We can not only work with abstractions, we can build machines to extend our ability to work with them (which we call computers). With the development of science in general, and mathematics and computer science in particular, we've thoroughly explored a lot of the intellectual territory available via symbol processing. If there is a bigger, better form of intelligence than symbol processing, I'd sure like to know what it is. If you're curious about this direction in the evolution of human intelligence, David Gelernter expresses some interesting ideas in his book *Mind in the Machine*.

   What makes hard problems hard is that they often aren't tractable to any form of analysis at all (e.g. weather prediction). One way to handle hard problems intelligently is to recognize when they're intractable, and then work on other problems.

5. *Changing IQ, impact on civilization*: If people were generally stupider, it would take much longer to get where we've gotten, and we'd be in worse shape when we got there. As an example, look at the Soviet Union, where people are effectively stupider because the social system there has punished intelligence and rewarded mediocrity. Now, that social system has changed, but the effects of 70 years of Pavlovian conditioning on several 100 million people won't go away overnight. We don't have a historical example for the reverse case. Presumably we would have progressed quicker, with a better quality of life.

6. *Memory capacity of brain*: To compute the proper value, first guess the average number of state bits in one neuron. Next, multiply by 15 billion. Guess the average number of bits required to represent a "memory." Divide the result from step 2 by the result from step 3. Do not attempt to use more than eight significant figures.

7. *$5000 for 100 IQ points?* When it's as common as Jiffy Lube, I'll think about it.

Don W. is a science fiction author from Austin, Texas, with an interest in the history of phrenology, the hereditarian IQ theory, monogeny, and polygeny. He comments:

1. *What is genius?* Genius is the desire to act upon what is already created or conditioned, coupled with the ability to do so and the strength of will to do so regardless of social constraints. Genius is despised close up, but the further you are removed from it in space and time the more attractive it seems. Genius is a curse from a dark god to hurt the regular patterns of the natural world. Genius is what ultimately transforms and is transformed by the matrix of existing thought and natural law. It is always sought for, but seldom found. It brings vast happiness to those not possessed of it.

2. *HyperIQ person's effect on humanity:* We couldn't measure such a high IQ, but assuming that we did have an individual with that cognitive ability, he would be treated as a freak unless he had equally high abilities of creativity, memory, and will-to-power. If he had those powers, he would be likely to conceal his powers (perhaps by living several simultaneous lives) and could possibly change society sufficiently to see the results of his work during his lifetime. But then how do you know he's not already here?

3. *HyperIQ person's world-view:* We have no science of mind. Chemistry is a science. If you walk from classroom to classroom you'll see the same periodic chart. Get on a plane and fly to Nairobi, New Delhi, Mexico City. Same chart. Walk from instructor to instructor in a university and you'll find that the Freudian can't agree with the Jungian. A hyperIQ individual would *have* to develop a real workable science of mind in order to survive. If you're going to spend your time cruising down the highway at 200 mph, you need a deep understanding of road physics. The hyperIQ individual would need to possess a great understanding of the mind, of thought and emotion, and, above all, of time since his perception of duration would be even more maddening than ours.

4. *Effect of superIQ person on humanity:* If it weren't for the limitations of the brain, we wouldn't be driven to expand ourselves, to develop our sense of self. The brain is a limiting device for the psyche.

Without the limiting effect of my brain, I couldn't be answering this survey—being too overwhelmed by the many possible tracks of thought and the barrage of sense data. Why would such an individual *want* to interact with humanity? Do you have any desire to try and make chimps better? Do you want to go down to the monkey house and try to teach them sign language? Of course not. Some hyperIQ might enjoy people as pets or lab animals.

5. *Changing IQ, impact on civilization*: Without genius there would be no paintings on cave walls, no burials of the dead, no new ways to tap flint, no names for the stars. (Indeed the influence of the night sky on mankind's psyche is no doubt one of the true sources of genius.) With IQ 100 as the top of the line, Sherwood Schwartz, creator of *Gilligan's Island*, would be our Shakespeare, and I shudder to think who would be our Sherwood Schwartz. If IQs were 40 points higher, our problem would be finding someone to do the dullest work. I suspect that there would be more tyranny as we forced the most mind-numbing (but necessary) work on those capable of greater things. Of course we do this anyway, which is why we have the spiritual and cultural problems we have. By the way, genius and obsessive–compulsive disorder individuals have a surface similarity in that both are driven to do what is beyond the understanding and customs of the world. The genius must change what is already created or conditioned. He or she is thus both a destroyer of the existent and a creator of the new. Both Shiva and Brahma. The obsessive–compulsive must maintain the order of things, no matter what the cost in time, effort, or outside abuse. The obsessive–compulsive must help the sun rise; the genius will make a new sun.

6. *Memory capacity of brain*: We already remember a great deal less than our ancestor's ancestors. Where is the man who can recite the complete *Odyssey*? Who among us knows the whole *Vedas*? I can't even recite my own poems that I read last night. All of our memory is external in papers and machines. But it is not wise to gain more memory, but to learn to gain better ones. This is what we need in an age of information whiteout. How horrible to know the theme song of "My Mother, the Car" (which I unfortunately do), and yet not a single poem by Eliot complete (which I don't).

7. *$5000 for 100 IQ points?* 100 point increase—Yes. 1000—No, because I couldn't stand all you zombies . . .

Ronald C. is a psychotherapist and admissions coordinator for a jail and hospital in Atlanta, Georgia. One of his interests are "the myths and manufacture of 'illnesses' to explain socioeconomic problems." He comments:

> Words like "talent," "interest" or "inclination" may be more appropriate than the word "genius." IQ is a bit bothersome. "IQ" is real and important, but I tend to challenge its specific meaning. Can we predict the height of a pyramid by measuring the base of the pyramid? Unfortunately, there are many factors involved in creativity, educability and the application of a basic IQ. Also there are psychological and environmental factors that shape people. In spite of Newton's genius he thought his research into the Book of Daniel and the Bible were highly important—or more important than his more famous endeavors. In regard to obsessive–compulsive behavior, a stereotypical obsessive–compulsive genius is found in a person like Tesla . . . but my hunch is that there are magnificent obsessions wrongly stigmatized as character flaws or chemical imbalances. Obsession is very much similar to or the same thing as addiction. Tesla was apparently fascinated by electricity at an early age, fascinated by the static of the fur of his pet cat, and the way snowballs radiated light when smacked against a wall.

Mark A. is technical director at Computer Artworks Ltd in London. His main interest is in artificial life. He comments:

1. *What is genius?* Random variation in brain structure produces individuals who are better or worse at given tasks. Genius, in any particular area, is the high percentile of this curve, often at the expense of other areas.
2. *HyperIQ person's effect on humanity*: If an individual with an IQ of 2280 could be implemented via human morphology, "human" may not be an applicable term. I think "artificial" organisms may feasibly reach this target. This emphasizes the point that a hyperIQ is unlikely to be considered "one of us." How would a population of pygmy chimps be affected by the presence of a human? The lower forms would be taught techniques that they are capable of understanding. The higher form's more abstract conceptualizations may be above the threshold and would remain a solitary pastime.

3. *HyperIQ person's world-view*: The very things to which this question refers, are, by definition, unknown. If you know what you don't know, you're halfway there.

4. *Effect of superIQ person on humanity*: DNA base counts are an even more misleading metric than IQ scores.

5. *Changing IQ, impact on civilization*: If there were never geniuses, there would be no microwave ovens or cable TV. We can understand what it would be like if human IQs were never above 100 by attending a soccer match in the UK. I would guess a WWF match, or a monster truck rally would be approximate U.S. equivalents. If IQs were always above 40, humanity would have been less warlike. A 40 IQ point median shift would not make a great deal of difference to the limits of our knowledge.

6. *Memory capacity of brain*: The brain stores terabytes—"large, but not *that* large." A person who had a brain injury started to remember every detail of life, becoming confused about past and present. Some of the redundancy of our brains, and its ability to recover from injury, is no longer needed in modern society.

7. *$5000 for 100 IQ points?* I would pay in an instant. Make it $5,000,000 and you would still have a taker, no question.

Hugh L. works at NASA Ames Research Center, California. He said:

1. *What is genius?* Dean Simonton's books address this subject.

2. *HyperIQ person's effect on humanity*: There is no way to create and validate an IQ test in this range. Ms. vos Savant is presumably $(228 - 100)/15$ sd's above the mean, which makes her one-in-I-forget. Which IQ test has been validated in that range, anyway?

3. *HyperIQ person's world-view*: As currently defined, a hyperIQ person could infer the rules used to create sequences in things like Raven's Progressive Matrices very quickly. Does this suggest the presence of some other level of knowledge? A hyperIQ person would be a brilliant mathematician capable of quickly creating new definitions which would generalize and simplify large areas of mathematics. Someone like Gauss, Hilbert, or von Neumann, for example. Intelligent though she may be, we obviously don't have valid tests in these upper ranges, since Ms. vos Savant did not demonstrate significantly high mathematical prowess with her book on Fermat's Last Theorem.

4. *Effect of superIQ person on humanity*: I don't see anything in your questions about "social intelligence," "niceness," etc. I suspect this is an equally significant question.

5. *Changing IQ, impact on civilization*: If IQ were clamped at 100, then mathematics could not have developed significantly, and therefore we would be stuck at the level of technology of the ancient Egyptians. Agrarian.

6. *$5000 for 100 IQ points?* I would pay for 50 points. The question isn't really meaningful beyond the validated ranges of tests. When you are one-in-five-billion, it gets difficult to define "IQ."

Mike H., a computer programmer at IBM Poughkeepsie and with an interest in cerebellum structure, comments on the last question:

$5000 for 100 IQ points? Do you take VISA?

Glenn K. who works at IBM in Fishkill, New York comments on the brain's memory capacity:

Memory capacity is probably too huge to guess at. However, anybody who spends time in a foreign country juggling various languages can tell you that there's an obvious limit on the size of the active vocabulary that's not very far above its actual size. As you add new words in one language, words in other languages seem to get booted out into more remote parts of memory to make room. For me, the hardest part of learning has always been memorization. Learning new ways to relate facts is easy, but increasing the region of memory devoted to those facts is a slow and painful process.

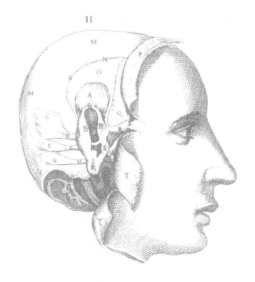

FINALE

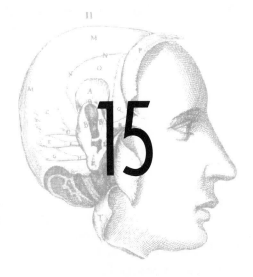

EPILOGUE

THERE ARE 10,000 WAYS TO GET TO ORIGINALITY.
SOME PEOPLE JUST HAVE INCREDIBLE
IMAGINATIONS. THAT DOESN'T MEAN THEY HAVE A
MENTAL DISORDER.

KAY R. JAMISON, M.D., 1993

WHO SO WOULD BE A MAN MUST BE A
NONCONFORMIST.

RALPH WALDO EMERSON

IT'S NOT ALWAYS EASY TO BE ECCENTRIC.

MARK TWAIN

You are in your museum of brains. Formalin's pungent odor fills your nose. In the distance, the humming of a heart–lung machine reminds you of the chanting of monks.

In your hand is the wet brain of the shyest of all scientists—the Honorable Henry Cavendish. This 18th-century scientist went beyond shyness, beyond ordinary social phobia. To ensure privacy, he developed an elaborate communication system of letter boxes and double doors in his house.

How could this pinkish-grey bulk in your hands have evolved from apelike ancestors? How is a brain that evolved for hunting and avoiding sabertooths able to have discovered two fundamental principles of electricity (Coulomb's and Ohm's laws) and realized that water is composed of hydrogen and oxygen?

You poke your finger into his spongy, convoluted fissures. Yes, Cavendish was shy, but not a burrower like William Harvey, the scientist who discovered blood circulation. Harvey built dark subterranean chambers in which to think. A slight smile comes to your lips as you recall Duke Bentinck-Scott who, like Cavendish, issued orders that no servants were to speak to him and used letter boxes on the door to his room. This molelike noble built underground tunnels, 15 miles in length, and an immense suite of subterranean rooms. The Duke emerged from his mansion only at night, following a little lantern carried by a servant ordered to stay forty yards away.

You pry apart Cavendish's lobes—so convoluted that you have the compulsion to trace their deep wrinkles with your finger. Yet his lobes had once been smooth when he was a fetus. Yes, all human brains start out smooth and gradually wrinkle like maturing mangos.

An adult gorilla's lobes are slightly smoother than an adult human's, a rat's still smoother, and reptile's still less convoluted. As we evolve, we wrinkle.

Cavendish never loved, was not entirely human. If a drug could have been given to Cavendish to increase his love and decrease his shyness, who knows what experiments he would have *failed* to conduct? Instead, he could have lived his wealthy life in marital bliss in some lavish castle with little time or will to carry out such demanding, exacting experiments. If Anafranil were available in the 1700s, would the state of modern science be retarded a hundred years? Or would obsessive–compulsive disorder inhibitors have freed all these geniuses from the prison of their minds, allowing them to soar to new heights on the wings of chemical angels? How will future scientific development be affected when the obsessive–compulsive disorder geniuses are eliminated from the world? Will we have made the individual happier at the expense of the universe?

Epilogue

It is time to leave the museum of brains. You look out a window and gaze up at the incredible lamp of stars.

Before you leave, you place Cavendish on the floor in the center of an immense array of brains that have taken humanity to new heights. Geoffrey Pyke's brain is now the freshest in your collection, although perhaps damaged by the large number of sleeping pills he took to kill himself. At least you were lucky to have arrived before his trochlear nerves were damaged. (They control the nerve that moves the eyes.)

You pop a few Anafranils into your mouth, and lock the door. You look back at your museum. On the door, engraved on an oak plank, is a motto from an obscure 20th-century eccentric:[1]

> "Blessed are the cracked,
> for they shall let in the light."

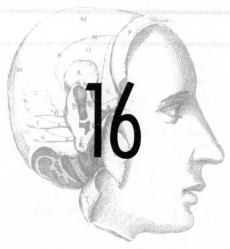

A TOUCH OF MADNESS

MEN OUGHT TO KNOW THAT FROM NOTHING ELSE
BUT THE BRAIN COME JOYS, DELIGHTS, LAUGHTER
AND SPORTS, AND SORROWS, GRIEFS, DESPONDENCY
AND LAMENTATIONS. AND BY THIS, IN A SPECIAL
MANNER, WE ACQUIRE WISDOM AND KNOWLEDGE,
AND SEE AND HEAR, AND KNOW WHAT ARE FOUL
AND WHAT ARE FAIR, WHAT ARE BAD AND WHAT
GOOD, WHAT ARE SWEET AND WHAT UNSAVORY . . .
AND BY THE SAME ORGAN WE BECOME MAD AND
DELIRIOUS, AND FEARS AND TERRORS ASSAIL US,
SOME BY NIGHT AND SOME BY DAY, AND DREAMS
AND UNTIMELY WANDERINGS ...

HIPPOCRATES

GREAT WITS ARE SURE TO MADNESS NEAR ALLIED.

JOHN DRYDEN

HOW I SELECTED "INFLUENTIAL" INDIVIDUALS FOR THIS BOOK

THERE WAS NEVER A GENIUS WITHOUT A TINCTURE
OF MADNESS.

ARISTOTLE

(ANOTHER TRANSLATION OF PRE-PREFACE QUOTE)

In selecting the influential, obsessive geniuses for this book, I purposefully included individuals scattered over two centuries, from Samuel Johnson (b 1709) to Geoffrey Pyke (b 1894). (Ted Kaczynski [b 1942], the Unabomber, does not quite fit this profile because his mathematical genius did not have a chance to flourish and be truly *influential* before he went off into the woods, but I include him here as a modern example of an obsessive genius with the potential for influence.) In determining who was influential, I did not simply choose the most famous or prestigious figures in science and philosophy. Nor did I confine my list of obsessive–compulsives to people who affected only the present state of science and philosophy. Influence on *past* generations was taken equally into account. I also considered the influence of their inventions and accomplishments on *future* generations and events. For example, it is certain that AC transmission of electricity will still be important decades from now, and the contributions of Tesla and Heaviside will therefore continue to affect the daily lives of our children. I have *not* attempted to place the strange geniuses in order of relative influence on the history of science, although such an exercise would provide lively classroom discussion and debate.

Why were there no women in my list of influential people with obsessive–compulsive disorder? For one thing, individuals in my list had both the talent and the opportunity to exert influence. Throughout history, women have usually been denied such opportunities, and my omission of females is simply a reflection of this situation.

I urge you to experiment by composing your own list of influential people with obsessive–compulsive disorder, which will probably include people not discussed in this book. I hope such an exercise will give you a new perspective on the history of science and reduce prejudice against people who live on the fringe of sanity but who nevertheless benefit humankind as much as, or maybe even more than, "normal" scientists and philosophers. (Again, because of Ted Kaczynski's killing spree, his overall effect on society is probably quite negative, but one wonders what he might have achieved if his genius were not cowled in violence. None of the other obsessive geniuses in this book had such a violent streak.)

WHY WERE "STRANGE BRAINS" BORN IN THE 18TH AND 19TH CENTURY?

IT COULD WELL BE THE LOW-KEYED TOLERANCE, THAT HAS ALMOST COME TO BE TAKEN FOR GRANTED IN THE BRITISH ISLES, THAT PROVIDE THE FERTILE GROUND UPON WHICH ECCENTRICITY MAY FLOURISH.

DAVID WEEKS AND KATE WARD
ECCENTRICS, 1988

GOOD SENSE IS THE ABSENCE OF EVERY STRONG PASSION, AND ONLY MEN OF STRONG PASSION CAN BE GREAT.

COMTE DE MIRABEAU

If you have come away with anything from this book, it is that being detached from the world is not solely a characteristic of the mystic. Obsessive–compulsive disorder scientists, through their determination, have harnessed insightful and practical visions.

Why were many of the strange geniuses in this book, as adults, virtually immune to stress that society places on nonconformists? Perhaps they excelled because the public in the 18th and 19th centuries was tolerant of their eccentric geniuses. Today, I am sure much of this tolerance would be transformed into hostility and hospitalization (or perhaps today these people would be living on the streets or in subway tunnels). As a result, we may have lost, or driven underground, potentially great thinkers. Are we paying too high a price in our attempts to fit in with our neighbors? Traditionally, eccentricity has not been distasteful. For example, many aboriginal cultures (e.g., native Americans) have been quite tolerant of sexual and mental nonconformity.

Victorian age people almost showed *affection* toward eccentrics. On the other hand, today, people who behave like Heaviside are homeless derelicts. Who would be modern candidates for the eccentric people in Jules Verne or H. G. Wells stories? Would the constraints of our modern society have forced Heaviside-like scientists to appear more normal? Would we recognize a Heaviside without his weird behavior? Isaac

A Touch of Madness

Asimov had his huge sideburns; Albert Einstein had his unruly hair; Richard Feynman had his bars, brothels, and bongos. Would a genius who looks and acts normal gain the same amount of attention (even serious attention) as someone who stands out by his special look or strange behavior? In the arts, the answer to this question is more clear. Many artists may feel compelled to cultivate their eccentricities to increase their public appeal. In Margaret Atwood's *The Cat's Eye*, a character makes reference to this idea:

> If I cut off my ear, would the market value of my paintings go up? Better still, stick my head in the oven, blow out my brains. What rich art collectors like to buy, among other things is a little vicarious craziness.

On the subject of Victorian geniuses, James Gleick in *Genius: The Life and Science of Richard Feynman* aptly comments:

> The romantics of the late 19th century saw powerful, liberating heroes, throwing off shackles, defying God and convention. They also saw a bent of mind that could turn fully pathological. Genius was linked with insanity—*was* insanity. That feeling of divine inspiration, the breath of revelation seemingly from without, actually came from within, where melancholy and madness twisted the brain.

GENIUS AND STRANGENESS

ECCENTRICITY IN SOCIETY IS ESSENTIAL FOR SUFFICIENT VARIETY TO ADAPT SUCCESSFULLY TO CHANGING CONDITIONS. ECCENTRICS MAY FAIL IN ANY SINGLE ENDEAVOR, BUT SOCIETY WINS BY THEIR EXAMPLES AND BY WHAT CAN BE SALVAGED FROM THE IDEAS, PROBLEMS AND QUESTIONS WHICH THEY RADIATE.

DAVID WEEKS AND KATE WARD
ECCENTRICS, 1988

THE HUMAN BRAIN DOES NOT GIVE UP ITS SECRETS WITHOUT A STRUGGLE. IN LIFE, IT HAS THE LOOK AND FEEL OF SOFT WHITE CHEESE. IT IS LACED

Strange Brains and Genius

WITH BLOOD VESSELS, SO THAT WHEN CUT IN
SURGERY, IT RESEMBLES CLUMPY YOGURT WITH
STRAWBERRY SAUCE. IT PULSES; YOU CAN SEE THE
HEART BEAT IN IT. OTHERWISE THE LIVING BRAIN
REVEALS LITTLE.

EVE LaPLANTE
SEIZED

Many geniuses through history have had peculiarities of one sort or another, or as Turin University professor Cesare Lomobroso once said, "Genius is often associated with anomalies in that organ which is the source of its glory." There are many theories about this association. For example, mental disease may cause individuals to overcompensate through constant creative activity. For others, the mental disease may actually "cause" creative ideas. In studying mad geniuses, we find that many have had a sense of physical vulnerability and the existence of a psychological "unease." (Perhaps this unease keeps individuals on edge and serves as a source of creative tension.) Their works often bear a personal mark and a striving for dominance or power. Almost all "mad geniuses" have had an irreverence toward authority, and a self-sufficiency and independence. Many of the trend-setting scientists experienced both social and professional resistance to their ideas. Nikola Tesla was often not taken seriously when he proposed correct ideas. Alexander Fleming's revolutionary discoveries on antibiotics were met with apathy from his colleagues. Niels Bohr's doctoral thesis on the structure of the atom was turned down by his university (the work later won him the Nobel prize). Joseph Lister's advocacy of antisepsis was resisted by surgeons.

Geniuses did not only differ from the norm in intelligence, but also in some other mental or physical characteristic as well. For example, Alexander, Aristotle, Archimedes, Attila the Hun, John Hunter, William Blake, and St. Francis Xavier were all short (most less than five feet, two inches tall). Aesop, Giotto, St. Bernard, Erasmus, Newton, Adam Smith, Boyle, Pope, Nelson, and Wren all had a physical deformity of one kind or another. Aristotle, Aesop, Demosthenes, Virgil, Darwin, and Cavendish all had a stammer. Arnold Ludwig, M.D., whom I discuss at the end of this section, suggests that chronic physical ailments give some individuals a heightened sense of urgency to leave a mark on the world and achieve immortality through creative greatness.

A Touch of Madness

Despite these facts, I do not know if these abnormalities occur more often in influential geniuses than in the general population. On the other hand, all my research points to the fact that there are certain *trends* in the lives of great geniuses, many of whom were listed in the Introduction of this book. For example, most of the individuals in this book have unusually self-reverential writing and speech. In addition, very few geniuses have given rise to exceptional progeny—either because great geniuses were celibate, or if married didn't want children, or because the children of geniuses were rarely geniuses (a fact that would have disturbed Francis Galton, who believed in the inheritance of intelligence.) Francis Bacon once wrote, "The care of posterity is most in them that have no posterity." None of the great English poets (Shakespeare, Johnson, Milton, Pope, Dryden, Goldsmith, Shelley, Keats, or Addison) produced exceptional children.

Many eminent people have had at least one child take his or her life, and many have had children with serious mental problems. Robert Frost's daughter was committed to the state mental hospital and another daughter had a "nervous breakdown." One of Albert Einstein's children was diagnosed as schizophrenic. Ambrose Bierce's oldest son committed suicide and his other died of complications brought on by alcoholism at 27. Thomas Edison had two children who became alcoholics, one of whom committed suicide. Alfred Stieglitz's daughter was psychotic and committed to a mental institution. James Joyce's daughter was admitted to an asylum for schizophrenia. There are numerous other examples that indicate frequent problems of children of geniuses. Many of these children tried unsuccessfully to pursue careers similar to their eminent parents. It is not clear if this played a role in their mental problems.

Many geniuses were distinctly asexual and never married. Michelangelo, for example, said, "I have more than enough of a wife in my art."[1] Recall Nikola Tesla's remark, "I do not think you can name many great inventions that have been made by married men." In addition to many of the strange brains described in this book, other notable celibates include Leonardo, Newton, Kant, Beethoven, Galileo, Descartes, Spinoza, Florence Nightingale, Copernicus, Handel, Cavour, Flaubert, and Chateaubriand. Philosopher Kierkegaard was celibate and considered sexual relationships an abomination. Saint Paul preferred sexual abstinence. Tennyson was asexual, never having kissed a woman until, at age 41, he married a 37-year-old invalid with a spinal problem. When Flaubert's seizures started, he gave up sex for a year and then developed homosexual interests. The cause for this low level of sexual interest in many geniuses is unclear. It could reflect disinterest in or aversion toward sex, an overriding preoccupation with work, or simply the lack of suitable opportunities for romance. Whatever the case, these "hyposexual" individuals were sexually abstinent or ascetic, giving few signs of having had romantic or sexual involvements with members of the opposite sex.

Arnold Ludwig, Professor of Psychiatry at the University of Kentucky Medical Center, notes that there are marked differences in sexual orientation among various professions. In his book *The Price of Greatness*, he notes that the creative arts professions attract higher proportions of homosexuals and bisexuals, the sciences attract (or promote) a higher proportions of hyposexual (sexually abstemious) people, and the "enterprising" professions, such as politics, business, and the military attract the highest proportions of heterosexuals.

Havelock Ellis published *A Study of British Genius* in 1904 and found that most geniuses had relatively old parents. Fathers' average age was 36, with very few fathers under 30. Mothers' average age was 31, with hardly a single mother under 25 producing a British genius. (Ellis considered anyone a genius with more than three pages in the 66-volume *Dictionary of National Biography* with 30,000 entries.) Also, as I alluded to in the Introduction of this book, geniuses often lose one of their parents shortly after birth.

Most geniuses display their special minds from early ages. Mozart was playing the piano at three. Galton and Bentham knew Latin and Greek at the age four. The French philosopher/scientist Pierre Gassendi preached at age four and explained lunar movements at age seven. The Italian poet Torquato Tasso spoke at six months and studied grammar at three years. Scientist Lord Kelvin entered Glasgow University at ten. The future Cardinal Wolsely graduated from Oxford at fifteen. Swiss polymath Albrech von Haller preached at age three, and at eight made lexicons containing all Hebrew and Greek words in both Testaments.

On the other hand, many great geniuses performed poorly at school. Such "dunces" included Isaac Newton, The Duke of Wellington (to be), physicist James Maxwell, and British foreign minister Sir Edward Grey, among many others. Albert Einstein's teacher said that his "presence in the classroom is disruptive and affects the other pupils." Many influential geniuses dropped out of school. Oliver Heaviside went to school until age 16 and had no formal education afterwards. Thomas Edison, the Wright

Brothers, George Stephenson, Charles Dickens, Mark Twain, and Maxim Gorky did not complete high school. Hume was called extremely weak-minded, and (Cardinal) Wiseman was labeled dull and stupid. Presidents Washington, Lincoln, and Truman never went to college.

Many geniuses such as Heaviside and Bentham were driven by fathers who were obsessive or ambitious to the point of cruelty. Anthony Smith, author of *The Mind*, notes that if many geniuses had no ambition, it was often coerced in them. Many geniuses were educated at home because the schools could not properly educate such advanced pupils. What would have happened to the strange brains I've discussed in this book if they were born today and legally required to attend public schools? "In compelling everyone to attend school," Smith says, "and taking education away from home, the opportunity for eccentricity, of which genius is a part, was diminished." One stunning example of the inadequacy of schools to accommodate genius is mathematician Norbert Wiener. Wiener read at age three and a half and was bored at school, so his father removed him until Wiener entered high school at age nine and graduated from Tufts College at 14. Like Cavendish, Wiener's father provided him a laboratory in which to experiment. In his 1948 book *Cybernetics, or Control and Communication in the Animal and the Machine*, Wiener attempted to show how a theory based on feedback and other engineering concepts could explain the operations of machines and biological and social phenomena. He coined the word "cybernetics" based on the Greek term *kybernetes* or steersman.

Young geniuses are the hardest to keep busy and interested in school, but they learn quickly and are usually very fluent and grammatical. (Note that Einstein was an exception and slow to speak.) They also have a remarkable range of knowledge. Young geniuses ask unusual questions, are curious and imaginative, and have strong powers of concentration in areas in which they are interested. Their parents are diverse. (Note that Kant, Kepler, Gauss, Faraday, Burns, Pasteur, and Luther were the progeny of semiskilled or unskilled workers.)

Arnold M. Ludwig, a psychiatrist at the University of Kentucky Medical Center in Lexington and author of *The Price of Greatness: Resolving the Creativity and Madness Controversy* (1995), has found that about one-third of the eminent poets, musical performers, and fiction writers suffered from serious psychological problems as teenagers, a rate that exploded to three-quarters when these people became adults. Even prominent scientists, who appear to suffer less mental disease than artists, exhibit a sudden steep rise in suicide rate in old age. As I, too, have found, Ludwig suggests that great creative achievers usually display their special abilities as children, are self-reliant, and face physical trials early in life, such as a disability or illness.

GENIUS AND MENTAL DISEASE

SOMETIMES A PATH TO RELIGION, SEIZURES ARE ALSO AT TIMES A PATH TO ART. DOSTOEVSKY ACTUALLY GAVE EPILEPSY TO AT LEAST FOUR OF HIS CHARACTERS, BUT MOST WRITERS WITH TEMPORAL LOBE EPILEPSY DISGUISE THEIR SEIZURES IN THEIR WORK.

EVE LaPLANTE
SEIZED

THERE IS QUITE DEFINITELY SOMETHING OR OTHER DERANGED IN MY BRAIN.

VAN GOGH

To what extent did obsessive–compulsive disorder play a role in the formation of the geniuses described in this book? As stated in the Preface, most scientists do not exhibit bizarre behaviors, and most people with obsessive–compulsive disorders do not possess extraordinary creativity. Yet obsessive–compulsive disorder seems to play a role in a certain kind of genius who invents by collecting materials with which to constantly experiment.

In an attempt to answer the question regarding obsessive–compulsive disorder and inventive geniuses, let us first consider various forms of mental illness that appear to stimulate creativity. For example, we know from studies reported in the February 1995 issue of *Scientific American* that bipolar disorder stimulates creativity, particularly in great artists, writers, and composers. Recent studies also suggest a connection between obsessive–compulsive disorder and bipolar disorder, as evidenced by similarity of some symptoms and the increased rate of obsessive–compulsive disorder in patients with bipolar disorder and depressive illness, as well as in their families (see Robins et al., Bellodi et al., and Rocca et al.). We also know from the work of Eve LaPlante and others that temporal lobe epilepsy

(TLE) seems to play some role with creativity in people such as Dostoevsky, Flaubert, van Gogh, and Lewis Carroll. For example, Dostoevsky admitted that his own creative ecstasy preceded a TLE seizure. van Gogh's phenomenal artistic output in the late 1880s also resulted from TLE. In fact, TLE can foster profound insights and mystical experiences to such a degree that some patients prefer not to take anticonvulsive drugs that deprive them of their creativity and of their enjoyable paranormal or mystical experiences.

TLE affects only part of the brain and alters (but does not eliminate) consciousness. Through history, TLE has affected a legion of writers, artists, and religious leaders. van Gogh, a temporal lobe epileptic, is a perfect example of the link between brain anomaly and creativity. Eve LaPlante in her book *Seized* describes van Gogh's classical signs of TLE: hyperreligiosity (heightened interest in religion), hypergraphia (excessive writing or drawing), and hallucinations. TLE often triggers religious needs and experiences. For example, van Gogh once wrote, "I often feel a terrible need of—shall I say the word?—of religion. Then I go out at night to paint the stars." van Gogh's hyperreligiosity manifested itself when he punished himself by refusing to eat and wearing rags while preaching Christianity. He had mystical visions, including those of a resurrected Christ, and had frequent attacks of rage. Since childhood, van Gogh had periodic headaches, stomachaches, dizziness, and depression. One day while painting flowers, the right half of his visual field went dark, while the flowers in the left half turned upside down. While he was shaving at a mirror, a TLE seizure produced a disembodied voice commanding him to kill painter Gauguin. Instead, van Gogh cut off his own ear, wrapped it in a paper, and left it at a brothel as a keepsake for a prostitute who had once posed for him.

van Gogh's TLE created his maddening desire to paint while heightening some of his perceptions. He called his seizures "the storm within," and he realized that they were the source of his hallucinations, unprovoked feelings of anger, fear, confusion, and uncontrolled floods of early memories—reminiscent of Nikola Tesla's uncontrolled visions. In one typical seizure, van Gogh saw every room in the house where he had been born, "every path, every plant in the garden, the views in the fields roundabout . . . down to a magpie's nest in a tall acacia in the graveyard." In 1888 while in a hospital, van Gogh painted more expressively than ever before. His hundreds of paintings in the hospital were nearly half his lifetime output. As his mind deteriorated, he created some of the most important works in the history of art. Van Gogh remarked that whatever was destroying him also fueled his creativity and sensitivity. Sometimes he drew sketches almost against his will. Would his creativity have been diminished if he had had modern anticonvulsive drugs?

At the height of his epileptic attacks, van Gogh began to scream and undress in public. His entire body would convulse, and he twice attempted suicide by swallowing turpentine, oil paint, and kerosene. Herman A. Harmelink, a van Gogh expert, has suggested that van Gogh's condition was caused by slow poisoning resulting from his eating oil-based paints, which can cause paranoia, seizures, and hallucinations. Today, Harmelink and I both know artists who joke about the dangers of eating oil-based paints and consider taking the risk, if it would mean they could paint like van Gogh.

In addition to hyperreligiosity and hypergraphia, TLE sufferers, like van Gogh, frequently experience diminished or increased sexual appetite, extreme dependence on other people (stickiness), and explosive tempers. Many of these characteristics typify the more abnormal individuals in *Strange Brains and Genius*. Oliver Heaviside, for example, was dependent on others his entire life, and many of the strange brains (e.g., Tesla, Heaviside, Bentham) were uninterested in sex. Geoffrey Pyke and Oliver Heaviside had classic symptoms of hypergraphia. Heaviside suspected that epilepsy was the cause of his strange behavior as evidenced by his statement, "My nerve disturbance . . . may culminate in epilepsy some day [like my mother]." Sometimes Heaviside had "violent convulsions, body and brain, like a gymnotus [eel] in a passion."[2]

Another common symptom experienced by temporal lobe epileptics is *jamais vu* (the feeling of never having been in what should be a familiar place—the opposite of *deja vu*). Sensory seizure symptoms include paresthesias (pins and needles feeling on the skin), formication (feeling bugs crawling under the skin), and the temporary inability to feel pain. Some temporal lobe epileptics have olfactory hallucinations such as suddenly smelling burnt feces. Others have seizural orgasms.

TLE affects sexuality in weird ways. La Plante describes one patient who, during a seizure, had a feeling identical to what she felt while masturbating, after which she felt terror and saw visions of "two-headed beasts, elephants, rhinoceroses, and hippopotamuses running around in circles and making no noise." (The woman's seizures stopped after a right temporal lobectomy.) A man with TLE was aroused by the sight of a safety pin. In fact, over time, he could orgasm simply by *imagining* a pin. A concert violinist and mother of three was compelled to dress like a prostitute and prowl the streets of Boston in search of the "ugliest, dirtiest" person with whom to have sex. After treatment with the anticonvulsive drugs Tegretol and Dilantin, her seizures ended and she returned to her musical career. As with most people who have TLE, this woman was misdiagnosed as a psychotic individual.[3]

Some people with TLE are convinced their way is the only right way of doing things, and therefore they become hypercritical of people around them. For example, consider the case of a TLE person who constantly complained to a supermarket manager that the products were not arranged in alphabetical order. The same patient wrote frequent letters to the utility companies complaining about perceived billing errors—reminiscent of Oliver Heaviside's constant letter writing to utility companies.

The religious and other obsessions exhibited by temporal lobe epileptics are similar to those seen in obsessive–compulsive disorder patients. One TLE patient, for example, was compelled to read *The Death of Socrates* by Roman Catholic priest Romano Guardini over and over again, underlining sentences and penciling in remarks. His bookshelves were filled to overflowing with religious books. "Sometimes I'm convinced there is a life after death and sometimes not," he said. "The Jewish people had magnificent faith for millennia before they even started thinking about life after death."

The overlap of some TLE symptoms with obsessive–compulsive disorder and bipolar disorder is significant. For example, one TLE patient compulsively made lists of her possessions and repeatedly sketched the sites of her religious experiences. Another patient described her religious feelings in detail, each line of text alternating in red and blue colors. Many temporal lobe epileptics compulsively underline and doodle excessively. Hypergraphia is also frequently exhibited by people with bipolar disorder. Interestingly, repeated use of LSD or cocaine can produce several similar traits. LSD alters temporolimbic function, causing people to become religious, hyposexual, fearful, and paranoid.

As with obsessive–compulsive disorder, we all have temporal lobe epilepsy to some degree. Our behaviors may not be as extreme as most of the geniuses in this book, but we've all had the TLE-driven feeling that we're being watched, or of *deja vu*, the feeling that we've been in a particular place before. Some of us have the ability to generate minor TLE episodes and induce visions and hallucinations as we fall asleep. Whenever a TLE seizure occurs, individuals have a tendency to explain the unusual experience in terms they can understand. In older days, the visions, voices, and heightened feelings were explained in term of religion. Today some explain them as alien presences.

GENIUS, EPILEPSY, GOD, ALIEN ABDUCTION

WITH TLE, I SEE THINGS SLIGHTLY DIFFERENT THAN BEFORE. I
HAVE VISIONS AND IMAGES THAT NORMAL PEOPLE
DON'T HAVE. SOME OF MY SEIZURES ARE LIKE
ENTERING ANOTHER DIMENSION, THE CLOSEST TO
RELIGIOUS OR SPIRITUAL FEELINGS I'VE EVER HAD.
EPILEPSY HAS GIVEN ME A RARE VISION AND
INSIGHT INTO MYSELF, AND SOMETIMES BEYOND
MYSELF, AND IT HAS PLAYED TO MY CREATIVE SIDE.
WITHOUT TLE, I WOULD NOT HAVE BEGUN TO
SCULPT.

A FEMALE TEMPORAL LOBE EPILEPTIC

TLE patients often believe that they are *controlled* from the outside, either by God, or by alien creatures from outer space. Eve LaPlante in *Seized* suggests that best-selling author Whitley Strieber has TLE. In 1987 Strieber wrote the book *Communion*, which described his abduction by 3½-foot aliens with two dark holes for eyes. Strieber[4] exhibits various symptoms of TLE: *jamais vu*, formiculation, vivid smells, hallucinations, rapid heartbeats, the sensation of rising and falling, and partial amnesia. Magnetic resonance imaging of Strieber's brain reveals "occasional punctate foci of high signal intensity" in his left temporoparietal region, which is suggestive of scarring that could lead to TLE. Strieber incorporated his occasional memory lapses and periods of altered consciousness into his "nonfiction" book *Communion*, which received a one-million-dollar advance from publishers. TLE probably accounts for a significant percent of out-of-body UFO abduction experiences. In fact, a significant number of abductees feel mild epileptic-like symptoms in advance of an "abduction." For example, some abductees feel heat on one side of their faces, hear a ringing in their ears, and see flashes of light prior to an abduction. Others report a cessation of sound and feeling or an overwhelming feeling of apprehension.

Alien abduction stories tell us about the workings of the mind. Michael Persinger, a neuroscientist at Laurentian University in Sudbury, Ontario, found that people with frequent bursts of electrical activity in their temporal lobes report sensations of flying, floating, or leaving the body, as well as other mystical experiences. By applying magnetic fields to the brain, he can induce odd mental experiences—possibly by causing bursts of firing in the temporal lobes. For example, he has made people feel as if two alien hands grabbed their shoulders and distorted their legs when he applied magnetic fields to their brains. However, some psychiatrists, such as John Mack, do not accept the TLE theory of abduction because no one has proven that abductees generally have excessive bursts of electrical activity in the brain's temporal lobes.

LaPlante and other TLE experts suspect that our modern fascination with ESP, out-of-the-body travel, past-life regression, and other paranormal phenomena may be the result of mild, undiagnosed TLE. Strieber, who has received many letters from people having similar UFO abduction experiences, notes:

> It's a terribly important and fundamental human experience—perceptions that come from the level of mind that isn't interrupted by the rational structures that animate most of our thought. It's a kind of memory, a form of perception, or a mechanism of consciousness, something inexplicable that the mind attaches an explanation to, probably the same thing that caused people to believe in the old gods and myths, in angels, resurrection, and even UFOs today. It probably starts in the human mind.

Streiber continues this line of thinking in *Transformation*, his 1988 sequel to *Communion*:

> It may be, that what happened to Mohammed in his cave and to Christ in Egypt, to Buddha in his youth and to all our great prophets and seers, was an exalted version of the same humble experience that causes a flying saucer to traverse the sky or a visitor to appear in a bedroom . . .

Unfortunately, most physicians wouldn't know what to do if a person with TLE showed up on their doorsteps. Neuropsychologist Paul Spires explains, "At most hospitals today, if you show up in the emergency room complaining that you feel panic, you saw a vision of the Virgin Mary dripping blood, and you smelled a dead mouse [all TLE symptoms], you're not going to get diagnosed as epileptic. You're going to be called schizophrenic and you're going to be sent straight to psychiatrist."[5]

TLE and obsessive–compulsive disorder have changed the course of civilization. LaPlante and others speculate that the mystical religious experiences of many of the great prophets were induced by TLE because the historical writings describe classic TLE symptoms. The religious prophets most often thought to have had epilepsy are Mohammad, Moses, and St. Paul. Even Dostoevsky thought it was obvious that Mohammad's visions of God were triggered by epilepsy. "Mohammad assures us in this Koran that he had seen Paradise," Dostoevsky notes. "He did not lie. He had veritably been in Paradise in an attack of epilepsy, from which he suffered as I do."

When Mohammad first had his visions of God, he felt oppressed, smothered, as if his breath were being squeezed from his chest. Later he heard a voice calling his name, but when he turned to find the source of the voice, no one was there. The local Christians, Jews, and pagan Arabs called him insane. Legend had it that Mohammad was born with excess fluid around his brain,[6] and that as a child he had fits. When he was five years old he told his foster parents, "Two men in white raiment came and threw me down and opened up my belly and searched inside for I don't know what." This description is startling in similarity to the alien abduction experience described by people with TLE such as Whitley Streiber.

Note that the overriding emotion experienced by Mohammed, Moses, and St. Paul during their religious visions was not one of rapture and joy but rather of fear. In 1300 B.C. when Moses heard the voice of God from a burning bush, Moses hid his face and was frightened. Luke and Paul both agreed that Paul suffered from an unknown "illness" or "bodily weakness" which Paul called his "thorn in the flesh." Famous biblical commentators have attributed this to either migraine headaches or epilepsy. Paul did once have malaria, which involves a high fever that can damage the brain. Other psychologists have noted that Moses, Flaubert, St. Paul, and Dostevesky were all famous for their rages and may have had TLE.

Psychologist William James, in *The Varieties of Religious Experience*, has argued that religious states are not less profound simply because they can be induced by mental anomalies:

Even more perhaps than other kinds of genius, religious leaders have been subject to abnormal psychical visitations. Invariably they have been creatures of exalted emotional sensitivity liable to obsessions and fixed ideas; and frequently they have fallen into trances, heard voices, seen visions, and presented all sorts of peculiarities which are ordinarily classed as pathological. Often, moreover, these pathological features have helped to give them their religious authority and influence. To plead the organic causation of a religious state of mind in refutation of its claim to possess superior spiritual value, is quite illogical and arbitrary (because) none of our thoughts and feelings, not even our scientific doctrines, not even our *dis*-beliefs, could retain any value as revelations of the truth, for every one of them without exception flows from the state of the possessor's body at the time. Saint Paul certainly once had an epileptoid, if not an epileptic, seizure, but there is not a single one of our states of mind, high or low, healthy or morbid, that has not some organic processes as its condition.

(Recall that several of the strange brains in this book, such as Galton, had "religious" visions of a Christlike figure.)

Recently, several TLE nuns have provided further evidence for TLE being at the root of many mystical religious experiences. For example, one former nun "apprehended" God in TLE seizures and described the experience:

Suddenly everything comes together in a moment—everything adds up, and you're flooded with a sense of joy, and you're just about to grasp it, and then you lose it and you crawl in to an attack. It's easy to see how, in a prescientific age, an epileptic or any temporal lobe fringe experience like that could be thought to be God Himself.

Even Ezekiel in the Old Testament had a TLE-like vision reminiscent of modern UFO reports:

And I looked, and behold, a whirlwind came out of the north, a great cloud, and a fire infolding itself, and a brightness was about it, and out of the midst thereof as the color of amber, out of the midst of the fire. . . . Also out of the midst thereof, came the likeness of four living creatures. And this was their appearance, they had the likeness of a man. And every one had four faces, and every one had four wings. And their feet were straight feet; and the sole of their feet was like the sole of a calf's foot; and they sparkled like the color of burnished brass.

Eve LaPlante is just one of a growing number of writers and researchers who believe in TLE-induced religious experiences. For example, Professor Michael Persinger from Ontario does research on the neurophysiology of religious feelings and believes that spiritual experiences come from altered electrical activity in the brain such as produced by TLE seizures. David Bear from Harvard Medical School believes that "a temporal lobe focus in superior individuals (like van Gogh, Dostoevsky, Mohammad, St. Paul, and Moses) may spark an extraordinary search for the entity we alternatively call truth or beauty." Religion is sometimes our interpretation of altered temporolimbic electrical activity. This is not to demean the mystical experience because TLE personalities have accomplished great things. Of course, TLE does not always produce beneficial results, as evidenced by the high percentage of TLE in violent criminals. Eve LaPlante in *Seized* aptly sums up the growing evidence linking TLE and creativity:

> Hidden or diagnosed, admitted or unknown, the mental states that occur in TLE seizures are more than simply neurological symptoms. In people like Tennyson, Saint Paul, and van Gogh these states may have provided material for religion and art. People with TLE, whether or not they know the physiological cause of their seizures, often incorporate their symptoms into poems, stories and myths. And the disorder does more than provide the stuff of religious experience and creative work. TLE is associated with personality change even when seizures are not occurring; it amplifies the very traits that draw people to religion and art.

MORE ON EPILEPTIC GENIUSES

IF WE COULD PROMOTE MORE ECCENTRICITY, OR ALLOW MORE VARIATIONS ON THE NORM STILL WITH DUE DIGNITY OR AT LEAST SOCIAL SAFENESS, THEN MENTAL ILLNESS TOO MIGHT BE LESS INTOLERABLE, LESS PAINFUL, LESS FULL OF STIGMA.

DAVID WEEKS AND KATE WARD
ECCENTRICS, 1988

A Touch of Madness

GENIUS IS A SYMPTOM OF HEREDITARY
DEGENERATION OF THE EPILEPTOID VARIETY.
CESARE LOMBROSO, 19TH-CENTURY PHYSICIAN

From the age of 22, French writer Flaubert had TLE seizures that began with a feeling of impending doom and of being transported to another dimension. He moaned, had a rush of memories, hallucinated, foamed at the mouth, moved his right arm automatically, fell into a ten-minute trance, and vomited. (Guy de Maupassant, the apprentice to Flaubert, had seizure-invoked hallucinations resulting from brain damage caused by syphilis.) Recall that Oliver Heaviside, too, had sudden, inexplicable feelings of "great anxiety, impending calamity, and despondency, and an aching desire for I know not what." Geoffrey Pyke "felt as epileptics must do when they are about to have an attack: a condition of generalized tenseness." Pyke often felt as if a "revelation" were about to occur. He would concentrate on the horizon and in another breath or two his mind was quite clear and he felt as content as "people do when they have received what they imagine is a religious revelation."

Hypergraphia has constructively been used by temporal lobe epileptics. For example, best-selling science fiction writer Philip K. Dick, who wrote the book on which the movie *Total Recall* is based, probably had epilepsy and hypergraphia. From age 15, he suffered from auditory and visual hallucinations that he interpreted as signs from God. He also had macropsia and micropsia (where stationary objects like chairs appear to enlarge or shrink) and depersonalization. Many of Dick's symptoms found their way into his books in the characters who had hallucinatory experiences. Dick published his first short story at age 13 writing at incredible speeds, and later produced thousands of pages of handwritten journals in addition to his 41 books.

Victorian poet and dramatist Alfred Lord Tennyson, along with his father, uncle, several brothers, cousin, and grandfather, had TLE. Tennyson wrote about his waking trances:

All at once, out of the intensity of the consciousness of individuality, the individuality itself seemed to dissolve and fade away into boundless being; and this not a confused state, but the clearest of the clearest, the surest of the surest, the weirdest of the weirdest, utterly beyond words. Sometimes a great and sudden sadness would come over me, and I would wander away beneath the stars.

Tennyson once had a vision of the entire inhabitants of London "lying horizontally a hundred years hence." He said, "The world seemed dead around and myself only alive. This might have been the state described by St. Paul."

If Tennyson and the other TLE artists and writers were alive today, we could perform several simple tests to determine if TLE were the cause of their strange brains. The simplest and most ingenious of these diagnostic methods is the "clock test." When TLE patients are asked to draw clocks, they very often produce clocks with excessive attention to details. For example, sometimes TLE people are so precise that they place the brand name on the clock and include all tick marks for minutes. I would like to see those who have had an alien abduction experience take the epilepsy clock test. As far as I know, such a test has not been applied.

What would have happened if the clock test were available to Tennyson and colleagues? Even if they were diagnosed as epileptic, they would not have had useful treatments. Luckily Tennyson, Flaubert, and van Gogh avoided the most painful 19th-century "treatments" for TLE, including prolonged application of a hot iron to the head, castration, or clitoridectomy. If today's anticonvulsant drugs were available to them, they might have been spared some of the debilitating TLE symptoms, but sacrificed some of their creative drive.

Even the great 19th-century poet Lord Byron had what he called "savage moods" where he bounced from periods of utter despondency to passionate, irrational urges. Although not epileptic, his volatile personality seemed to set off sparks of poetic imagination. Byron wrote that his affliction allowed him "from woe to wring overwhelming eloquence."

WHAT DOES IT ALL MEAN?

CREATIVITY AND GENIUS FEED OFF MENTAL TURMOIL. THE ANCIENT GREEKS, FOR INSTANCE, BELIEVED IN DIVINE FORMS OF MADNESS THAT INSPIRED MORTALS' EXTRAORDINARY CREATIVE ACTS.

BRUCE BOWER
SCIENCE NEWS, 1995

THE MANIC-DEPRESSIVE TEMPERAMENT IS, IN A BIOLOGICAL SENSE, AN ALERT, SENSITIVE SYSTEM

THAT REACTS STRONGLY AND SWIFTLY. THIS IS
IMPORTANT FOR CREATIVITY IN THE ARTS, THE
SCIENCES, AND LEADERSHIP.

KAY R. JAMISON, M.D., 1993

ECCENTRICS ARE THE HUMAN FACE OF CHAOS.

JEREMY GLUCK, 1995
NEW SCIENTIST

Genius and insanity are intertwined.[7] Professor Kay Redfield Jamison in her article "Manic-Depressive Illness and Creativity" (February 1995, *Scientific American*), clearly demonstrates that established artists have a significantly high incidence of bipolar disorder (manic depression) or major depression. (Bipolar disorder is a genetic disease where patients oscillate between depression and hyperactive euphoria.) Established artists and writers experience up to 18 times the rate of suicide seen in the general population, 10 times the rate of depression, and 10–20 times the rate of bipolar disorder. During periods of mania, patients have sharpened and unusually creative thinking, increased productivity, original thinking, expansive thoughts, and grandiose moods. They can overcome writing blocks, generate new ideas, and have better performances. People with bipolar disorder rhyme more often and use alliteration more often than unaffected individuals. They also use idiosyncratic words three times as often as control subjects and can list synonyms more rapidly than normal. As with TLE, some patients with bipolar disorder stop taking their medications because the drugs can dampen their emotional and perceptual range as well as their general intellect. However, in the future, we will worry less about the loss of creativity in eccentric geniuses because we will have found the genes responsible for the debilitating side of the afflictions, which should allow us to dampen the problematic symptoms while allowing individuals to retain their creative spark.

Jamison, a psychiatrist at Johns Hopkins University School of Medicine in Baltimore and author of *Touched with Fire: Manic-Depressive Illness and the Artistic Temperament* (1993), suggests that mania can be conducive to creativity because it allows afflicted individuals to work long hours without sleep, to focus intensely, to experience a variety of emotions, and to make bold assertions without fear of the consequences. Bipolars take risks and are open to contradictions. Jamison's descriptions about the effects of bipolar disease are sometimes poetic and mystical. In 1995, Jamison revealed in her memoir *An Unquiet Mind* that she, herself, has bipolar disorder. She describes a preternatural awareness of surroundings, a sensitivity to the environment that is more animal-like than human. She says, "You can't predict what you'll be like tomorrow. I think the moods of bipolarity mirror the natural world, which is so seasonal and fluid. It's a dangerous amphibious sort of existence."

Ruth Richards, a psychiatrist at the University of California, San Francisco, has found high levels of creativity both in people with mild mood swings and in healthy relatives of people with bipolar disorder. Alcoholism, depression, and mild mania occur often in mothers, fathers, and siblings of eminent individuals. Richards theorizes that genes involved in bipolar disorder have persisted in the human population because mild forms of the condition have conferred survival advantages: creativity, increased levels of work, and adaptability to a changing world.

Despite the fact that, as a group, artists display a much higher level of mental illness than do their creative counterparts in more structured occupations, not everyone agrees there is a causal link. For example, Arnold Ludwig, author of *The Price of Greatness: Resolving the Creativity and Madness Controversy*, believes that creative people who are mentally ill find themselves, almost by default, in the arts rather than in business or the sciences—fields that require rationality, persistence, and levelheadedness. He argues that the alternating moods of energetic enthusiasm and paralyzing depression that often accompany bipolar illness would preclude scientific achievement but not artistic achievement. And for cultural reasons, artists have little reason to curve flamboyant behavior because it actually contributes to their notoriety. Other clinicians believe that mental illness makes sufferers feel like outsiders—which can help motivate people to become artists. Creative greatness requires great dedication, to the point where relationships with people can be destroyed, which can be a price of being a genius.

It probably seems counterintuitive to many of you that obsessive–compulsive disorder could convey certain advantages to scientists and inventors. Perhaps the unlikely association conjures up simplistic notions of the "mad genius." No one wants to reduce the beauty and original thinking of these great people to a genetic flaw or disorder. But it is my opinion from researching the literature on TLE, obsessive–compulsive disorder, and bipolar disorder for this book that abnormal brain activity clearly plays a role in creativity, particularly in invention, writing, art, and music. However, the idea that behavior is controlled by brain states is not profound and should not surprise you. For example, even shyness has been proven to be an inborn personality trait, rather than an acquired one. (The brain structures dealing with shyness probably include the amygdala and hippocampus, which control heart rate and release of stress hormones.) Morality and religiosity also have a large organic component, as evidenced by intense hypermoralism and hyperreligiosity in those with TLE. Of course, there is a *range* of manifestations of brain disorders, and even in the ordinary person, the urge to create or be spiritual may be anatomically based. (I have decreasing use for Freudian psychology and believe that extraordinary advances in genetics, psychopharmacology, and neurobiology move modern psychiatric thought toward a more biological perspective.)

TLE, obsessive–compulsive disorder, and bipolar disorder all share certain symptoms with one another and with other disorders that manifest themselves in strange, inexplicable behavior. Some physicians believe that bipolar disorder and panic disorder may be seizure-based since anticonvulsant drugs sometimes help these patients. Schizophrenia is often associated with left-hemisphere epileptic lesions while bipolar disorder and depression are more common in patients with right-hemisphere lesions. I agree with the growing number of psychiatrists who believe that most major mental illnesses are caused by anomalies in brain structures that are influenced by biochemistry, neuroendocrine, and genetics and not by environmental factors such as parental upbringing. I find this concept is difficult for my friends to accept. Most people want to believe that their behaviors are volitional, under their control. Eve LaPlante concurs when she notes that mental disorders "challenge our assumption that we can control who we are and calls into question our conventional notions of credit and blame."

Would psychoactive drugs have decreased the productivity of the individuals we've explored? If their work were an expression of compulsiveness, drugs would have tempered that expression. On the other hand, drugs may have removed the shackles of their bizarre lifestyles. Obsessive–compulsive disorder allows geniuses to focus with extreme concentration on a subject, but, as a result, some of these individuals may have some difficulty in determining what are the most important questions for which to seek answers.

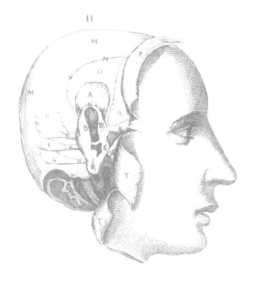

APPENDIXES

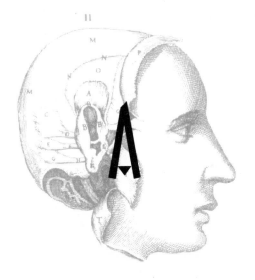

RUNNERS-UP LIST

IN AN ERA WHEN HUMAN BEINGS SEEM TYPECAST BY
THEIR CULTURE OR GENES, ECCENTRICS ARE A
REFRESHING REMINDER OF EVERYONE'S INTRINSIC
UNIQUENESS. BY HEEDLESSLY FLOUTING NORMS OF
BEHAVIOR THAT MOST OF US NEVER QUESTION, THEY
REMIND US HOW MUCH OF OUR LIBERTY WE
FORFEIT WITHOUT THOUGHT, AND HOW GREAT OUR
ABILITY IS, IN FACT, TO FORGE OUR OWN IDENTITIES
AND SHAPE OUR OWN LIVES.

DAVID WEEKS AND KATE WARD
ECCENTRICS, 1988

FEW OF US ARE NOT IN SOME WAY INFIRM, OR EVEN
DISEASED; AND OUR VERY INFIRMITIES HELP US
UNEXPECTEDLY.

WILLIAM JAMES
THE VARIETIES OF RELIGIOUS EXPERIENCE

considered numerous bizarre and brilliant individuals for inclusion in *Strange Brains and Genius*. The following individuals nearly became subjects for chapters in this book but were not included because they either did not have quite enough genius or did not display significantly aberrant behavior. To be the subject of a chapter, I required that the individual be either a scientist/inventor or philosopher, have the potential to be influential, be a genius, and have significantly strange behaviors.

THE ITINERANT FROM BUDAPEST

> MATHEMATICS IS THE ONLY INFINITE HUMAN ACTIVITY. IT IS CONCEIVABLE THAT HUMANITY COULD EVENTUALLY LEARN EVERYTHING IN PHYSICS OR BIOLOGY. BUT HUMANITY CERTAINLY WON'T EVER BE ABLE TO FIND OUT EVERYTHING IN MATHEMATICS, BECAUSE THE SUBJECT IS INFINITE. NUMBERS THEMSELVES ARE INFINITE.
>
> PAUL ERDÖS

> ERDÖS' DRIVING FORCE WAS HIS DESIRE TO UNDERSTAND AND TO KNOW. YOU COULD THINK OF IT AS ERDÖS' MAGNIFICENT OBSESSION. IT DETERMINED EVERYTHING IN HIS LIFE.
>
> RONALD GRAHAM, AT&T RESEARCH

Dr. Paul Erdös was a legendary mathematician who was so devoted to math that he lived as a wanderer with no home and no job. He kept working until 1996 when he died from a heart attack at the age of 83 while attending a mathematics meeting in Warsaw. Erdös (pronounced AIR-dosh) was one of the greatest, most eccentric, and most original mathematicians of all time. His passion was to both pose and solve difficult problems in number theory (the study of properties of integers) and other areas, and he founded the field of discrete mathematics, which is the foundation of computer science. He was also one of the most prolific mathematicians in history, with more than 1500 papers to his name.

Erdös is often remembered as being stooped and slight, and wearing socks and sandals. In order to pursue his mathematical obsession, he stripped himself of all the usual burdens of daily life—finding a place to live, driving a car, paying income taxes, buying groceries, and writing checks—and relied on his friends to look after him. "Property is nuisance," he said.

Because he had no interest in anything but numbers, his name is not well known outside the mathematics community. He wrote no best-selling books and showed little interest in worldly success and personal comfort. In fact, he lived out of a suitcase (which he never learned to pack) for much of his adult life. He gave away all the money he made from lecturing and mathematical prizes in order to help mathematics students or to offer cash prizes for solving problems he had posed. Erdös left behind only $25,000 when he died.

Sexual pleasure revolted him; even an accidental touch by anyone made him feel uncomfortable. Like many others in this book, Erdös never married or had a family. He often referred to husbands as "slaves" and wives as "bosses." Yet he was not a hermit like Kaczynski or Cavendish. In fact, Erdös hated to be alone, and almost never was; he loved to attend conferences and enjoyed the attention of mathematicians. His main aim in life was "to do mathematics; to prove and conjecture."

Concentrating fully on mathematics, Erdös traveled from meeting to meeting, carrying a half-empty suitcase and staying with mathematicians wherever he went. Ronald Graham of AT&T Research often handled Erdös' money for him, setting aside an "Erdös room" in his house for helping to manage Erdös' life. In fact, many of Erdös' colleagues took care of him, lending him money, feeding him, buying him clothes and even doing his taxes. In return, he inundated them with brilliant ideas and challenges—with problems to be solved and masterful ways of attacking them.

Dr. Joel H. Spencer, a mathematician at New York University's Courant Institute of Mathematical Sciences, once noted: "Erdös had an ability to inspire. He would take people who already had talent, that already had some success, and just take them to an entirely new level. His world of mathematics became the world we all entered."

Erdös was born into a Hungarian-Jewish family in Budapest in 1913, the only surviving child of parents who were mathematics teachers. (His two sisters, who died of scarlet fever, were considered even brighter than he was.) At the age of three he was amusing guests by multiplying three-digit numbers in his head, and he discovered negative numbers for himself the same year when he subtracted 250 degrees from 100 degrees and came up with 150 degrees below zero. He also delighted people by asking them how old they were and telling them how many seconds they had lived. A few years later, he amused himself by solving problems he had invented, like how long it would take for a train to travel to the sun.

When his father was captured in a Russian offensive against the Austro-Hungarian armies and sent to Siberia for six years, his mother removed him from school, which she was convinced was full of germs. Perhaps because Erdös' older sisters died of scarlet fever a few days before he was born, his mother became very protective of him. In fact, Erdös' mother was forever pampering him, and it is said that he never even buttered his own toast until he was 21 years old.

Erdös was educated mostly at home until 1930, when he entered the Piter Pazmany University in Budapest where he was soon at the center of a small group of outstanding young Jewish mathematicians. He practically completed his doctorate as a second-year undergraduate. Erdös received his doctorate in mathematics from the University of Budapest, and in 1934 came to Manchester on a postdoctoral fellowship.

By the time he finished at Manchester in the late 1930s it became clear that it would be an act of suicide for a Jew to return to Hungary. (Most members of his family who remained in Hungary were killed during the war.) Therefore, in 1938 Erdös sailed for the United States, where he was to stay for the next decade. During his first year at the Institute for Advanced Study in Princeton, he wrote ground-breaking papers that founded probabilistic number theory. He solved important problems in approximation theory and dimension theory. When his fellowship at the Institute was not renewed, he started his wanderings, with longer stays at the University of Pennsylvania, Notre Dame, Purdue, and Stanford.

The great mathematical event of 1949 was an elementary proof of the Prime Number Theorem, given by Atle Selberg and Erdös. The result, which predicts the distribution of primes with some accuracy, was first proved in 1896 by sophisticated methods, and it had been thought that no elementary proof could be given. (A prime number is one that has no divisors other than itself and 1.) Erdös was only 20 when he discovered this elegant proof for the famous theorem in number theory. The theorem, Chebyshev's theorem, says that for each number greater than one, there is always at least one prime number between it and its double.

Selberg and Erdös agreed to publish their work in back-to-back papers in the same journal, explaining the work each had done and sharing the credit. But at the last minute Selberg (who, it was said, had overheard himself being slighted by colleagues) raced ahead with his proof and published first. The following year Selberg won the Fields Medal for his work. Erdös showed little concern with the competitive aspect of mathematics and was philosophical about the episode.

Starting in 1954, Erdös began to have problems with the American and Soviet authorities. He was invited to a conference in Amsterdam but on the way back into the United States was interrogated by immigration officials about his Soviet sympathies. Asked what he thought of Marx, he gave his typically innocent answer: "I'm not competent to judge, but no doubt he was a great man." Denied his re-entry visa, Erdös left and spent much of the 1950s in Israel. He was allowed back into the United States in the 1960s.

Erdös never saw the need to limit himself to one university. He needed no laboratory for his work, no library or equipment. Instead he traveled across America and Europe from one university and research center to the next, inspired by making new contacts. When he arrived in a new town he would present himself on the doorstep of the most prominent local mathematician and announce: "My brain is open." While a guest, Erdös would often work furiously for a few days and then leave when he had exhausted the ideas or patience of his host. (He was quite capable of falling asleep at the dinner table if the conversation was not mathematics.) He would end sessions with: "We'll continue tomorrow, if I live."

Although his research spanned a variety of areas of mathematics,[1] Erdös continued his interest in number theory for the rest of his life, posing and solving problems that were often simple to state but notoriously difficult to solve and that, like Chebyshev's theorem, involved relationships between numbers. Erdös believed that if one can state a problem in mathematics that is unsolved and over 100 years old, it is probably a problem in number theory.

Erdös, like many mathematicians, believed that mathematical truths are discovered, not invented. He mused about a Great Book in the Sky, maintained by God, that contained the most elegant proofs of every mathematical problem. He used to joke about what he might find if he could just have a glimpse of that book.

In 1976 mathematical physicist Stanislaw Ulam wrote in *Adventures of a Mathematician*:

> Erdös had been a true child prodigy, publishing his first results
> at the age of eighteen in number theory and in combinatorial

analysis. Being Jewish he had to leave Hungary, and as it turned out, this saved his live. In 1941 he was twenty-seven years old, homesick, unhappy, and constantly worried about the fate of his mother who remained in Hungary. . . . Erdös is somewhat below medium height, an extremely nervous and agitated person. . . . His eyes indicated he was always thinking about mathematics, a process interrupted only by his rather pessimistic statements on world affairs, politics, or human affairs in general, which he viewed darkly. . . . His peculiarities are so numerous it is impossible to describe them all. . . . Now over sixty, he has more than seven hundred papers to his credit.

It is commonly agreed that Erdös is the second most prolific mathematician of all times, being superseded only by Leonhard Euler, the great 18th-century mathematician whose name is spoken with awe in mathematical circles. In addition to his roughly 1500 published papers, another 50 or more are still to be posthumously published. (Erdös was still publishing a paper a week in his seventies.) Erdös undoubtedly had the greatest number of co-authors (around 500) among all the mathematicians of all times. He collaborated with so many mathematicians that the phenomenon of the "Erdös number" evolved. To have an Erdös number 1, a mathematician must have published a paper with Erdös. To have a number of 2, he or she must have published with someone who had published with Erdös, and so on. Four and a half thousand mathematicians have an Erdös number of 2.

Many mathematicians consider Erdös the greatest problem poser of all times. For him building a new theory—a primary ambition of many mathematicians—was never his main goal. He was more interested in asking the right questions, and then the theory grew out by itself like a sapling. Erdös also had a superb ability to know which question to ask from whom. He commonly had three or four people sitting in different corners of a room, as he walked from one to the other, making significant simultaneous progress with each of them on problems belonging to quite different areas of mathematics.

Erdös would always make use of whatever time was available to do mathematics. For example, he would bring notebooks to concerts and start solving problems. It was common for him to listen to music and do mathematics at the same time. According to legend, on a long train journey he wrote a joint paper with the conductor.

As mentioned, Erdös was also exceptionally generous. Most of his money earned from royalties and awards was given to needy friends, young mathematicians having financial troubles, and charity. Because of his simple lifestyle, Erdös had little need of money. He won the Wolf Prize in 1983, the most lucrative award for mathematicians, but kept only $720 of the $50,000 he had received. (The rest went to friends, relatives, colleagues, and a large part of it to endow a postdoctoral fellowship at the Technion [Haifa] to commemorate his mother.) Lecturing fees also went to worthy causes. The only time he required funds was when another mathematician solved a problem that Erdös had proposed but not been able to solve. From 1954 he had stimulated his colleagues by offering rewards of up to $10,000 for these problems.

During his lifetime Erdös had many close friends and faithful and cherished disciples, but he had the deepest emotional contact with his mother, who started to accompany Erdös on his incessant travels when she was in her mideighties and continued to do so till her death at the age of 91 in 1971. The death of his mother was an incredible blow to Erdös from which he never fully recovered. After she died he found solace in doing even more mathematics than before. Erdös launched into his work with greater vigor, regularly working a 19-hour day. He fueled his efforts almost entirely by coffee, caffeine tablets, and benzedrine. He looked more frail, gaunt, and unkempt than ever, and often wore his pajama top as a shirt.

Despite, or perhaps because of, his eccentricities, mathematicians revered him and found him inspiring to work with. He was regarded as the wit of the mathematical world, the one man able to produce short, clever solutions to problems on which others had suffered through pages of equations.

Once, while pondering his own death, Erdös remarked:

> My mother said, "Even you, Paul, can be in only one place at one time." Maybe soon I will be relieved of this disadvantage. Maybe, once I've left, I'll be able to be in many places at the same time. Maybe then I'll be able to collaborate with Archimedes and Euclid.

A favorite saying of his was that "every human activity, good or bad, must come to an end, except mathematics." He died as he wished to, before his powers were greatly diminished and while attending a mathematics conference.

THE BONGO PLAYER FROM FAR ROCKAWAY

FEYNMAN WAS . . . WILD IN LOVE, VEERING FROM
FRESHMAN MIXERS (WHERE WOMEN SIDLED AWAY
FROM THIS RUBBER-LEGGED DANCER CLAIMING TO
BE A SCIENTIST WHO HAD MADE THE ATOMIC BOMB)
TO BARS AND BROTHELS.

JAMES GLEICK
*GENIUS: THE LIFE AND TIMES OF RICHARD
FEYNMAN*

Richard Feynman (1918–1988), American theoretical physicist, was the most brilliant, influential, and iconoclastic figure in his field in the post-World War II era. Feynman remade quantum electrodynamics—the theory of the interaction between light and matter—and thus altered the way science understands the nature of waves and particles. He was co-awarded the Nobel prize for physics in 1965 for this work, which tied together all the varied phenomena at work in light, radio, electricity, and magnetism. Feynman received his B.S. at the Massachusetts Institute of Technology (MIT) in 1939 and his Ph.D. at Princeton University in 1942. During World War II, he worked on the atomic bomb project at Princeton University (1941–42) and Los Alamos, New Mexico (1942–45). In 1945, he became an associate professor of theoretical physics at Cornell University. In 1950, he became professor of theoretical physics at California Institute of Technology. For almost two decades he taught a course known as "Physics X" where undergraduates would gather to pose any scientific question they wished, and Feynman would improvise.

Feynman also contributed to the key equation for the efficiency of nuclear explosion, and he developed a theory of predetonation that measured the probability that a lump of uranium might explode too soon. He organized the world's first large-scale computing system, which consisted of electro-mechanical calculators and teams of women passing color-coded cards!

Born in the Far Rockaway section of New York City, Feynman was the descendant of Russian and Polish Jews who had emigrated to the United States late in the 19th century. His parents had great expectations for, and a profound impact on, young Feynman. For example, his father announced even before Feynman was born that his son would be a scientist. Even before baby Feynman was out of his high chair, his father brought and placed various tiles in front of Feynman's eyes to get him accustomed to looking for patterns. Like Einstein, Feynman was a late talker, but as he grew up he soon began to devour books. When his father brought him the *Encyclopaedia Britannica* at a young age, Feynman was unstoppable.

Like other electrical geniuses in this book, as a child he accumulated tubes sets and batteries and assembled with various gadgets: transformers, switches, coils, and sparking devices. As a teenager his passion for mathematics grew and he began to fill notebooks with formulas. He and his friends traded mathematical curiosities like baseball cards. In his high school yearbook, he was voted the "Mad Genius."

Biographer James Gleick notes in *Genius: The Life and Times of Richard Feynman*:

> Originality was his obsession. He had to create from first principles—a dangerous virtue that sometimes led to waste and failure. He had the cast of mind that often produces cranks and misfits: a willingness, even eagerness, to consider silly ideas and plunge down wrong alleys.

While not nearly as obsessive–compulsive as the others in this book, Feynman knew how to focus on the research problem needing his attention and force all distractions from his field of vision. Like Nikola Tesla, Feynman surrounded himself with an aura of myth, and spent a great deal of time generating anecdotes about himself. Like other strange brains we've discussed, Feynman's inspiration often came in flashes, and he always had a skepticism about official wisdom and an impatience with mediocrity.

As Feynman went through life, he explored all aspects with gusto. He learned to play the bongos, to give massages, and to pick up women in bars. He learned how to keep very accurate track of time in his head, how to open safes, how to mimic foreign languages, how to train dogs, how to persuade young women to disrobe, how to write Chinese, and how to make columns of ants march to his bidding. His bongo playing reflected his perpetual interest in rhythm. While at MIT he fell into the habit of bothering his roommates with the absentminded drumming of his fingers and a tapping staccato on walls and garbage pails. After Feynman finished his sophomore year at MIT he suffered from something akin to a nervous exhaustion. He soon recovered, and throughout his life practiced dream control and was able to dream the same dream again and again, with variations, at will.

Despite his obvious genius, Feynman's score on his school IQ test was merely a respectable 125. Yet Feynman was the classic genius, a mind breaking with standard methods and seeing the world new. He searched for principles that would be flexible enough so that he could adapt them to anything in the universe, and he chided his graduate students who would begin their work on a problem in the normal way, by checking what already had been done. When they did this, he told them they would give up chances to find something original.

Feynman was born a Jew but he was areligious and viewed science and religion as adversaries. Interestingly, in a study of Nobel laureates, one researcher found that Jewish persons made up 19 percent of the 286 Nobelists of all nationalities named up to 1972, a percentage many times greater than their representation in their various countries of origin. Some have suggested that this high percentage may partly reflect the predilection of Jews to go into scientific fields such as molecular biology and medicine that are eligible for the Nobel prize, rather than those such as botany, and zoology, which are not.[2]

James Gleick comments about the genius of scientists like Feynman: "Geniuses display an undeniable obsessiveness resembling, at times, monomania. Geniuses of certain kinds—mathematicians, chess players, computer programmers—seem, if not mad, at least lacking in the social skills most easily identified with sanity."

THE RECLUSE FROM SHREWSBURY

Charles Robert Darwin (1809–1882), the naturalist whose discovery of the theory of evolution by natural selection revolutionized biology, was born in Shrewsbury, England. His father had one of the largest medical practices outside of London, and his grandfather was the physician Erasmus Darwin, the author of *Zoonomia, or the Laws of Organic Life*. Darwin's mother died when he was eight years old, but otherwise he had a happy childhood.

As a boy, he was very interested in specimen collecting and chemical investigations, but he was an uninspired student at school. At age 16 he was sent to study medicine at the University of Edinburgh, where he was repelled by surgery performed without anesthetics. His father was disappointed by Darwin's disinterest in medicine and next sent him to the University of Cambridge in 1827 to study divinity. At the time, Darwin adhered to the conventional beliefs of the Church of England.

While at Cambridge, his older cousin William Darwin Fox (an entomologist who inspired Darwin's lifelong zeal for collecting beetles) began teaching Darwin about natural sciences. Others encouraged Darwin's excitement about science and confidence in his own abilities, and this started Darwin on his famous career and exploration of tropical islands.

While at sea, Darwin seemed to enjoy his journey, despite his seasickness. Adventurous, he seemed to relish danger and was sustained in the considerable discomfort by a lively curiosity. He wrote to one of his sisters:

> We have in truth the world before us. Think of the Andes; the luxuriant forest of the Guayquil; the islands of the South Sea & New South Wales. How many magnificent & characteristic views, how many & curious tribes of men we shall see. What fine opportunities for geology & for studying the infinite host of living beings: Is this not a prospect to keep up the most flagging spirit?

When Darwin returned from his voyage, he began drawing up lists of the benefits and drawbacks of marriage. He finally proposed to his first cousin Emma Wedgwood, whom he married in 1839. She brought fortune, devotion, and considerable domestic skills that enabled him to work in peace for the next 40 years. Newly married, the Darwins moved into a house in London, but within a few years Darwin's increasingly poor health prompted them to move to the country.

Charles and Emma Darwin had 10 children: two died in infancy and a third died at age 10. The surviving five sons went away to school. Three became distinguished scientists, and one a major in the royal army who was also an engineer and eugenicist. A fifth son was undistinguished, as were his sisters, who prepared at home to follow their mother into marriage.

Darwin was devoted to his wife and daughters but treated them as children. For example, Emma had to ask him for the only key to the drawers containing all the keys to cupboards and other locked depositories. Nevertheless, Darwin was offended by all expressions of real cruelty, and he was an ardent opponent of slavery.

His ideas about women seemed to shape his scientific insights. For example, Darwin believed that the young of both sexes resemble the adult female in most species and reasoned that males are more evolutionarily advanced than females. "The female is less eager than the male," he wrote. "She is coy," and when she chooses a mate, she looks for "not the male which is most attractive to her, but the one which is least distasteful."

Darwin's medical school experience had left him sympathetic to the popular antivivisectionist movement, but he admonished women for their involvement in it in an 1876 letter to *The Times of London*:

> Women, who from the tenderness of their hearts and from their profound ignorance are the most vehement opponents of such experiments, will I hope pause when they learn that a few such experiments performed under the influence of an anesthetic, have saved and will save through all future time thousands of women from a dreadful and lingering death.

Comfortable in English society, Darwin worried about alienating and offending people with his theory on evolution. He had inherited much of English society's conservatism, and his fear of ostracism was one reason he waited before publishing his theory. Darwin felt particularly bad about the pain his theories would inflict on his close friends and especially on Emma, all devout Christians for whom his theory was heresy.

Some believe that Darwin's worries about the effect of his theories on society caused him to develop physical pain. In fact, the once-adventurous young naturalist became a semi-invalid and hermit before his 40th year. Physical symptoms included painful flatulence, vomiting, insomnia, and palpitations. Although he was exposed to insects in South America and could possibly have caught Chagas or some other tropical disease, a careful analysis of his symptoms in the context of his activities points to psychogenic origins.

Darwin's illness has been the subject of extensive speculation. It all began within a year after returning from his famous voyage exploring the South American continent and the offshore islands, when he retreated to the country living life as a virtual recluse. What was wrong with him? In the January 8, 1997, *Journal of the American Medical Association*, psychiatrists Russell Noyes and radiologist Thomas Barloon concluded that Darwin had panic disorder.[3]

Aside from the physical symptoms such as heart palpitations, light-headedness, shortness of breath, nausea, and vomiting, he developed extreme sensations of fear that lasted until his death in 1882. He also suffered from severe anxiety and hysterical crying. Darwin spent a lot of time "treading on air and vision," which researchers say suggests feelings of depersonalization. He feared going out, suggesting agoraphobia[4] often seen with panic disorder. Over the years doctors have continued to puzzle over Darwin's disease and have suggested diagnoses such as parasitic disease, arsenic poisoning, depression, epilepsy, and inner ear disorder. But the researchers suggest that panic disorder is the most likely culprit.[5]

Like other strange brains in this book, there may have been compensations for Darwin's mental disabilities. He once wrote, "Ill health has annihilated several years of my life but has saved me from the distraction of society." Physicians Noyes and Barloon conclude that, "had it not been for this illness, his theory of evolution might not have become the all-consuming passion that produced *On the Origin of Species*."

THE HYPOCHONDRIAC FROM HOUSTON

Howard Hughes (1905–1976) was an inventor, manufacturer, aviator, and motion-picture producer who was famous for his aversion to publicity as well as for the uses to which he put his money.

Born in Houston, Texas, he later studied at the California Institute of Technology, Pasadena, and at the Rice Institute of Technology, Houston. Orphaned at 17, he quit school and took control of his father's Hughes Tool Company, Houston. In 1926 he moved to Hollywood, where he produced several movies.

In 1948 he bought a controlling interest in RKO Pictures Corporation, sold the shares in 1953, bought the whole company in 1954, selling it again in 1955. He remained chairman of the board until 1957.

In the field of aviation he founded the Hughes Aircraft Company, Culver City, California, and used the profits to finance the Howard Hughes Medical Institute. On September 12, 1935, in an airplane of his own design, he established the world's landplane speed record of 352.46 miles per hour. Two years later, in the same plane, he averaged 332 miles per hour in lowering the transcontinental flight-time record to 7 hours 28 minutes. In 1938, flying a Lockheed 14, he circled the Earth in a record 91 hours 14 minutes. In 1942 he began work on the design of an eight-engine, wooden flying boat intended to carry 750 passengers. In 1947 he piloted this machine on its one mile flight (its first and last flight).

Always an introvert, Hughes went into complete hiding in 1950. As the holder of 78 percent of the stock of Trans World Airlines (TWA), he refused to appear in court to answer antitrust charges and thus lost control of the organization by default. In 1966 he sold his TWA shares for more than $500,000,000. His strong desire for privacy and seclusion aroused great interest in his whereabouts. In his last years he abruptly moved his residence from one place to another (Bahamas, Nicaragua, Canada, England, Las Vegas, Mexico), arriving at each new destination unnoticed, taking special precautions to ensure total privacy in a luxury hotel, and rarely being seen by anyone except a few male aides. He often worked for days without sleep in a black-curtained room, and he became emaciated and deranged from the effects of a frugal diet and too many drugs. He died enroute when flying from Acapulco, Mexico, to Houston, Texas, to seek medical treatment.

Like Samuel Johnson, Richard Kirwan, and Oliver Heaviside, Howard Hughes had an extreme preoccupation with health that came to dominate his entire life. As an example, consider the instructions he gave to his staff for removing the hearing aid cord from the cabinet in which it was stored:

> First use 6 or 8 thickness of Kleenex pulled one at a time from the slot in touching the door knob to open the door to the bathroom. The door is to be left open so there will be no need to touch anything when leaving the bathroom. The same sheaf of Kleenex may be employed to turn on the spigots so as to obtain a good force of warm water. This Kleenex is then to be disposed of. The hands are to be washed with extreme care, far more thoroughly than they have ever been washed before, taking great pains that the hands do not touch the sides of the bowl, the spigots, or anything else in the process.

He required his aides to wear gloves when handing him objects and to write messages to him rather than spread germs by communicating verbally. He had his doors and windows sealed with tape to avoid dirt. He refused to touch doorknobs. (At the time, this cluster of behaviors was diagnosed as obsessive–compulsive order.)

When he became constipated, he would spend up to 26 hours in the bathroom. Sometimes he would urinate on the floor, but would not let his help clean it up, preferring instead to spread towels on the floor. Later, he started to urinate against the bathroom door. He began to spend hours cleaning his phone with tissues. The slightest change in his physical condition would cause him great anxiety, and he took all kinds of medicines to protect himself from disease and germs. (This cluster of behaviors was diagnosed as hypochondriasis.)

Despite his obsessions, he often could gather himself together to conduct business. As his condition deteriorated, though, he would spend more time in bed, rarely bathe, and let his finger and toe nails grow abnormally long. (This cluster of behaviors was diagnosed as psychosis.)

THE ROCK MAN FROM LATVIA

OH! HOW NEAR ARE GENIUS AND MADNESS
DENIS DIDEROT, FRENCH PHILOSOPHER

In the early 1920s, Edward Leedskalnin (four feet, eight inches tall) emigrated from Latvia to America where he began to undertake a strange outdoor sculpting hobby. The romantic breakup with his 16-year-old lover was the stimulus for his life's work of solitary construction in a remote part of Florida. Here he surrounded his property with huge coral block walls and filled the courtyard with bizarre monuments.

While working secretly for 30 years, and using parts mostly from an old Ford car, he sculpted the local rock into 12 rocking chairs, a map of Florida, astronomical instruments, various planets, and a 25-foot tall obelisk. Perhaps because of his broken romance, Leedskalnin sculpted two adjacent beds, and cradles for children. The main entrance was through a gateway blocked by a nine-ton coral slab that opened by a touch of a finger. Engineers still wonder how Leedskalnin found the balance point on which this massive stone pivots and also how Leedskalnin moved all the gigantic stone blocks. Everything he did was unobserved and done single-handedly.

While spending his solitary hours in his stone garden, Leedskalnin pondered on electricity and magnetism and conducted experiments with car batteries. He believed that modern science did not correctly understand such phenomena as plant growth and the motions of planets, which he believed were caused by the flow of magnetic particles having sight, taste, and other senses. His most interesting publication was *A Book in Every Home*, which gave various "practical" advice such as, "Everyone should be trained not to go out anywhere before someone else has examined them to see if everything is all right. It would save many people from unexpected embarrassment." He also advised people not to feed their daughters too much because they would get "too large." Parents were encouraged to teach their children not to smile because smiling leaves wrinkles. (Every other page was blank in *A Book in Every Home* so that if readers disagreed they could write their comments on alternate pages.)

In addition, Leedskalnin suggested that girls be virgins prior to marriage. Therefore, a girl's mother should have sex with the daughter's suitor in order to train the boy. "In case a girl's mamma thinks that there is a boy somewhere who needs experience," Leedsklanin writes, "then she, herself, could pose as an experimental station for that fresh boy to practice on and so save the girl. Nothing can hurt her any more. She has already gone through all the experience that can be gone through and so in her case would be all right."

The secret of how Leedskalnin manipulated the huge stones died with him in 1951. In the end, like many of the strange brains in this book, Leedskalnin lived on a very monotonous diet of only sardines and crackers. Eventually he forgot to eat at all and died of starvation.

THE SELF-TREPANARS FROM LONDON

BETWEEN THE PHYSIOLOGY OF THE MAN OF GENIUS,
THEREFORE, AND THE PATHOLOGY OF THE INSANE
THERE ARE MANY POINTS OF COINCIDENCE.

CESARE LOMBROSO, 1891

Sometime in the late 1960s, Amanda Fielding and Joey Mellen came to believe that their consciousness could be beneficially altered by liberating their brains from the imprisonment of their skulls. In order to test this hypothesis, they drilled holes in their skulls. (Drilling a hole in your own head is called "self-trepanation.") Joey Mellen in 1975 wrote the wonderful book *Bore Hole* with the opening sentence, "This is the story of how I came to drill a hole in my skull to get permanently high." Later in his book he described the effects of self-trepanation:

> I felt brilliant, god-like, able to understand everything. At the same time as being fascinated by the way I could see things as through a magnifying glass, I could hear all the sounds of the town outside the house as well as those inside, and each perception registered quite clearly, distinct from all the others though related to them, like the various instruments in an orchestra. Now I know what eternity meant. Time seemed to stop, and still everything was moving. I was ecstatic.

It all started when Mellen found a trepan instrument in a surgical store. The trepan looks like a hand-operated drill and has a central spike to hold the device in position. Around the spike a sharp cutting disc revolved. When the drilling is finished, a small circular disk of bone is removed, exposing the brain.

Mellen's first attempt at boring a hole into his skull didn't go very well. Before starting the operation, he administered a local anesthetic and took LSD to reduce his nervousness. Next he sliced open the skin to reveal the bare bone. The hardest part was to sink the central spike into position before drilling. Joey Mellen couldn't do it on his own.

The next time he tried to drill a hole, Mellen enlisted the help of his friend Amanda Fielding. With all her strength she plunged the spike into the skull, and it took hold. Mellen started to drill but fainted and had to be taken to the hospital. Horrified doctors said if he had drilled any deeper he would have injured the brain and killed himself.

Soon after this incident, Mellen tried again. The results are described in his book *Bore Hole*:

> After some time there was an ominous sounding schlurp and the sound of bubbling. I drew the trepan out and the gurgling continued. It sounded like air bubbles running under the skull as they were pressed out. I looked at the trepan and there was a bit of bone in it. At last! On closer inspection I saw that the disc of bone was much deeper on one side than on the other. Obviously the trepan had not been straight and had gone through at one point only. Then the piece of bone had snapped off and come out. I was reluctant to start drilling again for fear of damaging the brain membranes with the deep part while I was cutting through the rest, or of breaking off as splinter. If only I had had an electric drill it would have been so much simpler. Amanda was sure I was through. There seemed to be no other explanation of the scourging noises. I decided to call it a day. At that time I thought that any hole would do, no matter what size. I bandaged up my head and cleared away the mess.

To be sure he had actually penetrated the brain, Mellen tried again in 1970. This time he used an electric instrument which burned out after half an hour's work. A neighbor repaired it, and Mellen finally was able to break through to the brain. During the next few hours, he reached a state of tranquility which he says has been with him since the operation.

His companion Amanda Fielding also trepanned herself while recording the operation on film. Surgical scenes are interspersed with attractive motion studies of Amanda's pet pigeon. The soundtrack is soothing music. Like Mellen, Fielding has claimed to have achieved inner peace as a result of the experience. The part of the film that really shocks me is the scene in which blood spurts as she penetrates the skull, followed by her triumphant smile.

My most recent information suggests that Joey and Amanda live happy, normal lives together and have started a family. A photograph I have shows Joey, Amanda, and their son, Rock, outside their Chelsea gallery. Family friends claim Joey and Amanda are more relaxed and believe themselves to be satisfied with life since the self-trepanation.

THE CAVEMAN FROM FOLKESTONE

William Harvey (1578–1657) is the great English physician who discovered how the blood circulates and the function of the heart. Among his famous patients were two kings of England (James I and Charles I) as well as philosopher Francis Bacon. His monumental book *An Anatomical Treatise on the Movement of the Heart and Blood in Animals* has been called the most important book in the entire history of physiology. Harvey also did pioneering work in embryology.

Harvey was bizarre and brilliant. As a young man he wore a dagger and drew it at the slightest provocation. Like other strange brains in this book, Harvey had no children. Like Tesla and others, he had scotaphilia (love of the dark). He worked in the dark because he could think better, and he constructed underground caves beneath his house in Surrey for meditation. Harvey exhibited hypergraphia (excessive writing), and sometimes added repeating letters to the end of his words for no logical reason. For example, in one place in his notes he spelled "pig" as "piggg," and, as one Harvey biographer wrote, this is "an unusually liberal number of 'g's' even for 17th-century English." Other words also have strange repetitions of their last characters.

THE RAT LADY FROM EDINBURGH

Susanna Kennedy, the 18th-century Countess of Eglintoune, was a six-foot tall beauty with many admirers. Many of her beaus compared her to

Cleopatra or Helen of Troy. The secret of her beauty, she said, was that she washed her face daily in sow's milk and never used makeup.

When her husband asked for a divorce because Susanna had seven daughters but not a single son, she said she would consent if he returned to her everything she possessed when she first got married. He agreed, thinking she was referring to monetary assets. When she explained, "Return to me my youth, beauty, and virginity, and then dismiss me as you please," he never raised the subject again. Her husband died in 1729.

The Countess believed she had never received sufficient gratitude from humans. Her favorite companions, like Jeremy Bentham's, were rats, which she kept in large numbers. She would summon them to the dining room at mealtimes by banging on an oak panel. A dozen rats would respond to her request and join her at the table. After dinner, she signaled the rats to leave, at which point they left the dining room in an orderly manner and waited to be summoned again for their next repast.

OTHER RUNNERS-UP

During the writing of *Strange Brains and Genius*, many of my colleagues suggested various individuals for consideration as chapter subjects. Here are some other brilliant individuals considered to have been afflicted with temporal lobe epilepsy (TLE): Peter the Great, Alexander the Great, Moliere, Eugene Delacroix, Rasputin, August Strindberg, and actor Richard Burton. As I indicated previously, TLE predisposes geniuses to detect connections and transform vision into art.

Neurologist William Gordon Lennox and other psychologists consider the following geniuses epileptic: writers Petrarch, Tasso, and Charles Dickens; musicians Handel and Paganini; religious figures Saint Cecilia and Buddha; philosophers Socrates, Pascal, and Swedenborg; political leaders Julius Caesar, Cardinal Richelieu, and Napoleon; and mathematicians Pythagoras and Sir Isaac Newton.

Professor Kay Redfield Jamison views the following individuals as suffering from major depression or from bipolar disorder: Sylvia Plath, Walt Whitman, Cole Porter, Anne Sexton, Vincent van Gogh, Gustav Mahler, John Berryman, Edgar Alan Poe, Virginia Woolf, Herman Hesse, Mark Rothko, Mark Twain, Charles Mingus, Tennessee Williams, Georgia O'Keeffe, Ezra Pound, and Ernest Hemingway.

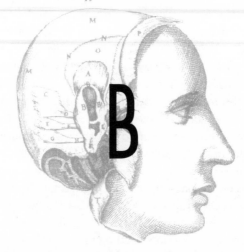

UPDATES AND BREAKTHROUGHS

IT'S BEEN SAID THAT GENIUS IS NEXT TO INSANITY.
WHEN YOU LOOK AT IT NEUROPHYSIOLOGICALLY,
THIS STATEMENT MIGHT HAVE SOME VALIDITY.
GENIUS APPEARS TO BE CHARACTERIZED BY THE
ABILITY TO DRAW MORE CORRELATIONS BETWEEN
DISTANT OR SEEMINGLY UNRELATED CONCEPTS
THAN A LESS-THAN-GENIUS WOULD BE ABLE TO DO.
SOME FORMS OF INSANITY RESULT FROM THE
EXISTENCE OF MENTAL CONNECTIONS THAT
"NORMAL" PEOPLE CONSIDER EXTRANEOUS OR
DETRIMENTAL. GENIUS AND INSANITY BEAR
NOTABLE SIMILARITIES.

ANDREW LUNDIN

SHE USED TO EAT CHOPS IN THE SMALL HOURS OF
THE MORNING AND SLEEP IN A HAT. ONCE, SHE
ARRIVED HOME AT SEVEN A.M. CARRYING A GATE.

> WHO AM I TO SAY THERE IS ANYTHING WRONG
> WITH HER?
>
> SATIRIST ALAN COREN
> *THE SANITY INSPECTOR*

While writing *Strange Brains and Genius*, I have monitored current scientific breakthroughs in understanding the biological basis for mental illness, talent, and behavior. After the short "Swallowers" section, I present a sampling of recent information on epilepsy, schizophrenia, and other related areas. Most information comes from news articles that have come across my desk during the writing of this book.

SWALLOWERS AND OTHERS

Chapter 10 on obsessive–compulsive disorder dealt with individuals who perform endless rituals that dominate their daily lives. For example, trichotillomaniacs continuously pull out their hair. Almost all people afflicted with obsessive–compulsive disorder recognize their problems as soon as they start and therefore often keep their embarrassing diseases hidden.

The most extreme case of compulsive swallowing is that of a 42-year-old woman who complained of a "slight abdominal pain" in 1927. Physicians removed 2533 objects from her stomach including 947 bent pins. In 1985, 212 objects were removed from a man whose stomach contents included: 53 toothbrushes, two razors, two telescopic aerials, and 150 handles of disposable razors.

On the flip side of the coin are those individuals who, for a variety of reasons, insert objects into their rectums. The medical journals list an astonishing array of objects: a bottle of pancake syrup, an axe handle, a nine-inch zucchini, a plastic spatula, a Coke bottle, an 11-inch carrot, an antenna rod, a 150-watt light bulb, 72 jeweler's saws, an apple, a frozen pig's tail, an 18-inch umbrella handle, two Vaseline jars, a teacup, an oilcan, a 6×5-inch tool box, a two-pound stone, a baby powder can, a peanut butter jar, a ball-point pen, baseballs, a sand-filled bicycle inner tube, sewing needles, a flashlight, a tobacco pouch, a turnip, a pair of eyeglasses, a hard-boiled egg, a carborundum grindstone, a suitcase key, tumblers and glasses, and a polyethylene waste trap from the U-bend of a sink. In 1955, one depressed man inserted a six-inch paper tube into his rectum, dropped in a lighted firecracker, and blew a hole in his anterior rectal wall. For a more comprehensive list of objects which people have voluntarily placed inside themselves, see: Adams, C. (1988) *More of The Straight Dope*. Ballantine: New York.

Another unusual manifestation of obsession and hypergraphia relates to compulsive diary writers. Perhaps the most famous case of very detailed diaries was reported in a *Seattle Times* feature that chronicled Robert Shields, 77, of Dayton, Washington, as the author of a 38 million-word personal diary. Filling 81 cardboard boxes, the diary allegedly covered the past 24 years of his life in five-minute increments. An example: July 25, 1993, 7 a.m.: "I cleaned out the tub and scraped my feet with my fingernails to remove layers of dead skin." 7:05 a.m.: "Passed a large, firm stool and a pint of urine. Used five sheets of papers."

THE ANATOMY OF OBSESSION

Recent research by Jeffrey Schwartz and his colleagues at the UCLA School of Medicine suggests that behavioral therapy for obsessive–compulsive disorder changes not only the behavior of people but also their brain chemistries. Typical therapy encourages people afflicted with obsessive–compulsive disorder to expose themselves to situations that provoke an obsessive response. By gradually increasing the duration of the exposure, the obsessive–compulsive behavior often diminishes. Before starting such therapy, Dr. Schwartz uses positron-emission tomography scans to examine key brain structures whose chemical activity is abnormal in obsessive–compulsive disorder patents. Using these medical imaging techniques, he finds that these brain structures have different chemistries and operate normally after successful behavioral therapy.

In particular, four anatomical regions are known to be involved with obsessive–compulsive disorder. The orbital cortex lies just above the eye sockets. Three other structures lie deep inside the brain—the caudate nucleus, the cingulate gyrus, and the thalamus. As an example, the orbital cortex is overactive in humans, and it is precisely this region that when damaged leads to obsessive–compulsive disorder in monkeys. In humans with obsessive–compulsive disorder, all four regions of the brain use glucose at very high and correlated rates, as if the four regions were working together to the detriment of the afflicted person. The fact that behavior therapy can reduce this high and correlated brain chemistry, even in adults, suggests that the brain is plastic much later in life than people had previously thought. (For further information: Glausiusz, J. (1996) The chemistry of obsession. *Discover*. June, 1996.)

GENETICS AND OBSESSIVE-COMPULSIVE DISORDER

Scientists have found that a variant of a specific gene may contribute to obsessive–compulsive disorder, at least in men. This genetic alteration reduces the production of the enzyme catechol-O-methyltransferase, or COMT, which helps terminate the action of the neurotransmitters dopamine and norepinephrine. Researchers at the Rockefeller University in New York suggest that it may be one of several "susceptibility" genes that pose a risk for developing obsessive–compulsive disorder.

The researchers focused their attention on the COMT gene because it has a known function and falls within a DNA segment that, when missing, is associated with symptoms of manic depression, schizophrenia, and obsessive–compulsive disorder. A variation of the COMT gene's usual sequence occurs in nearly half of the tested men suffering from obsessive–compulsive disorder. In contrast, only 1 in 10 women with obsessive–compulsive disorder displayed this variation, as did one in six of the men and women who displayed good mental health. Statistical analyses identified the COMT variant as a likely contributor to obsessive–compulsive disorder in men. (For further information: Bower, B. (1997) Gene may further obsessions, compulsions. *Science News*. May 3, 151: 269.)

SCHIZOPHRENIA AND CHEMISTRY

Schizophrenia, a mental disorder that disrupts thought and emotion, has joined many other major mental illnesses whose cause is biological with little or nothing to do with parental upbringing or family relations. In many cases of schizophrenia, nerve cells in the adult brain's prefrontal cortex lack messenger RNA molecules to carry out genetic instructions for creating an enzyme that makes the neurotransmitter gamma-aminobutyric acid (GABA). This brain malfunctioning in *adult* schizophrenics appears to result from a defect in *fetal* brain development. Because brain cells deprived of GABA resemble healthy cells, the brains of schizophrenics appear healthy. These findings may suggest an organic mechanism for the negative symptoms of schizophrenia: apathy, disorganized thoughts, and emotional flattening. (For further information: Bower, B. (1995) Schizophrenia: fetal roots for GABA loss. *Science News.* 147(16): 247.)

Another contributor to schizophrenia is an inadequate supply of a protein that enhances the flexibility of brain structures involved in learning and memory. In particular, neural cell adhesion molecules (NCAM) are severely reduced in the sea-horse-shaped hippocampus of schizophrenic brains. This NCAM shortage may originate during fetal development. (For further information: Molecular clue to schizophrenia (1995) *Science News.* 147: 202.)

SCHIZOPHRENIA AND GENETICS

Scientists have recently identified a gene that makes individuals susceptible to schizophrenia. In particular, several genes on chromosome 6 may trigger a chain of chemical reactions that results in some forms of this mental illness. (Scientists have found alterations in chromosome 6 in family members with schizophrenia.) A known mutation of one of the genes on chromosome 6 causes damage to the cerebellum, a portion of the brain involved in muscle coordination and some types of thinking. (For further information: Genetic hint to schizophrenia (1995) *Science News.* 147(19): 297. See also: Bower, B. (1995) Schizophrenia yields new gene clues. *Science News.* 148(19): 292.)

SCHIZOPHRENIA AND BRAIN STRUCTURE

Some of the most severe and intractable symptoms of schizophrenia— such as a deadening sense of apathy and social withdrawal—appear in youngsters displaying the smallest brains, as measured by brain-scanning devices. Small brain size may create a vulnerability to a number of mental disorders, which then combine with disturbances of brain development specific to schizophrenia. (For further information: Bower, B. (1996) Childhood clues to schizophrenia. *Science News.* 151: 40.)

Another study suggests that defects in a brain circuit can explain all the symptoms of schizophrenia. Crucial components of this circuit include the cerebellum, which may pace thinking efforts, the thalamus, which filters incoming information, and the prefrontal cortex, a center of complex thinking and judgment. Damage to the various parts of this circuit appears to account for the shifting spectrum of social, emotional, and thinking difficulties experienced by people suffering from schizophrenia. (For further information: Bower, B. (1996) Faulty circuit may trigger schizophrenia. *Science News*. 150(11): 164.)

EPILEPSY AND THE IMMUNE SYSTEM

A rare but serious form of childhood epilepsy is caused by antibodies attacking the brain. Rasmussen's encephalitis starts with a small seizure, such as a twitch of the eye or mouth, and develops into more pronounced seizures. The only reliable treatment is removal of the entire affected half of the child's brain. Researchers have recently found that children with this form of epilepsy have antibodies to the receptor through which glutamate, a brain neurotransmitter, stimulates nerve cells. When scientists filtered antibodies out of the blood of a nine-year-old girl with this form of epilepsy, the girl improved dramatically. Prior to treatment, she had ten major seizures a day, and lay in bed waiting for the next seizure. After treatment, she walked, talked, read, and wrote. (For further information: Barinaga, M. (1995) Antibodies linked to epilepsy. *Science*. 268: 262.)

READING DISABILITY, ATTENTION-DEFICIT DISORDER, AND GENETICS

Researchers suspect a genetic, immunological basis for reading disability (RD) and attention-deficit hyperactivity disorder (ADHD). The possible gene responsible for both RD and ADHD resides on chromosome 6 within a region concerned with producing substances that protect the body against viruses and bacteria. Decreased amounts of this substance have been found in ADHD children or in their mothers or both—and also in autistic children. The relationship between RD, ADHD, autism, and the immune system is an active area of research. (For further information: Warren, P. et al. (1995) Reading disability, attention-deficit hyperactivity disorder, and the immune system. *Science*. 268(5212): 786.)

A GENE IS LINKED TO AUTISM

Autism is a complex and cruel disorder. Although it is known to be largely genetic in origin, researchers have been unable to pin down its mechanism, partly because only about 10 in 10,000 are afflicted. In 1997, researchers found what appears to be the first link between autism and a specific gene that regulates serotonin, a key brain chemical. In particular, autistic children are significantly more likely to have a shortened form of the serotonin transporter gene. This gene codes for a protein that reabsorbs serotonin into the neuron that has released it. A shortened version would be expected to make more serotonin available for its receptor. Currently this finding is counterintuitive, because antidepressant drugs such as fluoxetine (Prozac) increase the availability of serotonin and often suppress autistic symptoms such as repetitive and ritualistic behaviors associated with aggression or anxiety. Perhaps the gene needs to act in concert with other (identified) genes to lead to autism. (For further information: Holden, C. (1997) A gene is linked to autism. *Science.* 276(5314): 905.)

MUSICAL ABILITY AND THE BRAIN

Scientists have found the region of the brain that allows some people to have perfect pitch—the ability to identify any musical note without comparison to a reference note, a talent displayed by Mozart and others. The "planum temporale," a region of the brain that processes sound, is far larger on the left side than on the right in professional musicians who have perfect pitch. While only 1 in 10,000 people have perfect pitch, this ability is relatively common among professional musicians. Perfect pitch also has an increased occurrence in autistic people where its incidence is 1 in 20. Individuals with Williams syndrome also have a similar increased incidence of perfect pitch. (Those afflicted with Williams syndrome have abnormally acute hearing, often resulting in pain on hearing only moderately loud sounds.) Perfect pitch usually manifests itself in people before the age of five. (For further information: Nowack, R. (1995) Brain center linked to perfect pitch. *Science.* 267: 616; Sacks, O. (1995) Musical ability. *Science.* 268(5211): 621.)

LANGUAGE AND THE BRAIN

Recent research suggests that a single network of brain regions allows people to speak both in their native language and a second language.

According to new brain scan studies, a language learned later in life is represented in the brain in the same region as a native language. (For further information: Brain's singular way with language (1995) *Science News*. 147: 202.)

VIRUS MAY TRIGGER DEPRESSION AND OBSESSIVE-COMPULSIVE DISORDER

The Borna disease virus (BDV), first reported 100 years ago to cause neurological diseases of horses in Borna, Germany, may also infect humans and trigger major depression and other severe mood disorders. Scientists find that markers for BDV coincide with acute episodes of mood disorders in humans, and the markers disappear during recovery. German investigators have found BDV genetic material in the blood cells of four patients hospitalized for major depression or obsessive–compulsive disorder. (For further information: Bower, B. (1995) Virus may trigger some mood disorders. *Science News*. 147(9): 132.)

ANATOMY AND MURDER

Brains of murderers often show less activity in the frontal cortex than brains of nonviolent people of the same age and sex. In one study of 22 murders, about 75 percent had low frontal activity, which is believed to regulate aggressive impulses. (For further information: Gibbs, W. (1995) Seeking the criminal element. *Scientific American*. 272(3): 101.)

DEPRESSED BABIES' BRAINS

Babies born to depressed mothers grow up with brains exhibiting unusual electrical activity. In particular, the infants' right frontal lobes display more electrical activity than the left. The cause is unknown, but previous work suggests that emotions linked to social engagement spark left frontal activity, whereas feelings associated with social withdrawal elevate right frontal activity. A similar electrical imbalance also exists in the depressed mothers' brains. (For further information: Sparks of inhibition. (1995) *Science News*. 147(21): 333.)

BIPOLAR DISORDER LINKED TO ABSENT DNA

Researchers have found that a severe form of manic depression, usually beginning by late childhood, afflicts many people suffering from a genetic disorder known as velo-cardio-facial syndrome (VCFS), the primary features of which include cleft palate, heart defects, and learning disabilities. VCF stems from the deletion of a specific segment of chromosome 22. A new study finds that nearly two-thirds of people with VCF suffer from bipolar disorder (manic depression). A gene known to be located in the deleted chromosome 22 region produces an enzyme that activates two chemical messengers, dopamine and norepinephrine. These substances have been implicated in bipolar disorder. (For further information: Bower, B. (1997) Manic depression linked to absent DNA. *Science News.* 151(1): 7.)

SEXUAL ABUSE SHRINKS BRAINS

Severe, repeated sexual abuse in childhood leads to significant reduction in the size of the hippocampus, a seahorse-shaped brain structure that controls memory. Researchers postulate that this damage may predispose people to experience dissociation and posttraumatic stress disorder (PTSD). The hippocampus is also small in Vietnam combat veterans suffering from PTSD.

Information on brain "shrinkage" comes from magnetic resonance imaging of the brains of women who suffered prolonged sexual abuse before age 15. (For further information: Bower, B. (1995) Child sex abuse leaves mark on brain. *Science News.* 147(22): 340.)

GENETICS AND HOMOSEXUALITY

Evidence continues to mount regarding the genetic underpinnings of homosexuality. Two independent studies suggest that a gene linked to male homosexuality resides on a small region (q28) of the X chromosome. Also, a significant number of homosexual brothers share the Xq28 marker, but homosexual men with heterosexual brothers usually fail to share this marker. (For further information: Holden, C. (1995) More on genes and homosexuality. *Science*. 268(5217): 1571. Also see: Travis, J. (1995) X chromosome again linked to homosexuality. *Science News*. 148(19): 295.)

IMAGINED PHYSICAL DEFECTS

Horrifying bumps blossom from behemoth bulging noses. Faces crawl with crimson spots and growths. Breasts and genitals shrink or enlarge in drastic ways. Mouths stretch from ear to ear like creatures from *The Outer Limits*.

People with "body dysmorphic disorder" (BDD) incorrectly perceive that something is drastically wrong with their faces and bodies. They continually check the mirror, undergo numerous plastic surgeries, and sink into years of isolation and despair.

It now appears that the same brain disturbances responsible for obsessive–compulsive disorder contribute to BDD, and the same serotonin-boosting drugs such as Prozac or Anafranil that help relieve the symptoms of obsessive–compulsive disorder also help people with BDD. BDD occurs frequently in combination with obsessive–compulsive disorder—nearly one in four people with obsessive–compulsive disorder suffer from BDD as well. (Obsessive–compulsive disorder afflicts about 2.4 million people in the United States at some time in their lives.) When Anafranil is given to people with BDD, their deformities actually appear to melt away, facial spots disappear, and horrifying body hair seems to fall out. These visual illusions support their assertions of the physical reality of their bodily defects.

There are also some significant differences between obsessive–compulsive disorder and BDD. People with BDD rarely realize that their perceptions are illogical and unfounded. They get little relief from looking in a mirror and often cannot function in public or at work. For example, one woman with BDD was so convinced that her nose was misshapen that repeated testimonials to the contrary by friends and plastic surgeons (and repeated plastic surgeries) could not change her conviction and endless despair. In contrast, people with obsessive–compulsive disorder usually realize that their behavior is strange, feel better after performing rituals, and function relatively well in public. (For further information: Bower, B. (1995) Deceptive appearances, imagined physical defects take an ugly personal toll. *Science News*. 148(3): 40.)

IMAGINED VOICES AND BRAIN STRUCTURE

Schizophrenics often hear disembodied voices uttering demeaning phrases such as, "You are worthless and stupid." These schizophrenics with auditory hallucinations have brain parts with reduced blood flow when the individuals are asked to imagine sentences in another person's voice. In contrast, normal individuals and nonhallucinators do not have reduced blood flow in these areas of the brain (the supplementary motor area and the left middle temporal gyrus) when asked to imagine sentences in another person's voice. (For further information: Bower, B. (1995) Brain scans seek roots of imagined voices. *Science News*. 148(11): 166.)

GENETIC BASIS FOR EXCITABLE PERSONALITY

Researchers have found a gene that participates in shaping a specific personality trait. In particular, a version of the D4 dopamine receptor gene appears frequently in people who report high levels of "novelty seeking." Such individuals enjoy exploring new environments, are excitable and quick-tempered, and seek out thrilling sensations. (For further information: Bower, B. (1996) Gene tied to excitable personality. *Science News*. 149(1): 4.)

OEDIPUS, SCHMOEDIPUS

Neurosis may have a genetic cause. Researchers have discovered two variations of a gene that affects the way brain cells respond to serotonin. The gene comes in a short and long version. The short version promotes fewer of the molecules that facilitate the reabsorption of the brain chemical serotonin. The long promotes the creation of more such molecules. Researchers found a significant correlation between the presence of the short gene and what Freud liked to call neurosis (fear, dread, worry, pessimism, suspiciousness, anxiety . . .) It's possible that this gene results from natural selection. While humans were evolving, a certain amount of fear and caution was a good thing. (For further information: Collins, J. (1996) Oedipus, Schmoedipus. *Time*. December: 74.)

GENE PAIR MAY INCITE OBESITY, DEPRESSION

Genetic influences on human obesity are being increasingly researched, and new studies suggest two genetic variations as contributing to the condition, as well as to depression and anxiety. Women who have certain chemical alterations in regions near the so-called human obesity (OB) gene appear substantially more likely to be obese than women with the more common form of this DNA. When these changes near the OB gene occur in combination with a version of the D2 dopamine receptor gene known as the A1 allele, the likelihood of obesity in young women rises even further. Controversial studies have linked the A1 allele to a broad range of substance abuse and related disorders, such as pathological gambling, that may be mediated by reward centers in the brain. (For further information: Bower, B. (1996) Gene pair may incite obesity, depression. *Science News*. 150: 181.)

IS CHOLESTEROL A MOOD-ALTERING LIPID?

In a series of studies over the past decade, researchers have found that people with low concentrations of cholesterol in their blood are more likely than average to attempt suicide. Two recent research papers rekindle interest in this possible association. In one recent study of men, the risk of suicide increases with low cholesterol concentration and in those men whose concentrations decline through time. In women, those with the greatest drop in cholesterol concentrations after delivering their babies are most likely to suffer postpartum depression. (For further information: Bower, B. (1996) Is cholesterol a mood-altering lipid? *Science News*. 150: 184.)

HAPPINESS AND DNA

According to recent research, our genes are primarily what determines our baseline level of happiness. While a person's sense of well-being changes in response to life's events, these are short-term fluctuations. In the long run, people tend to return to a set point, a level of well-being that varies from person to person and is largely genetically determined. The researchers' conclusions were based largely on studies of identical and fraternal twins. For example, identical twins, but not fraternal twins, show marked similarity to one another in self-ratings of happiness and contentment, regardless of their education, income, marital status, or religious commitment. (For further information: Happiness and DNA. (1996) *Science*. 272: 1591.)

Other studies of global happiness suggest that in 37 of 43 countries for which nationally representative samples are available, average ratings of well-being and happiness fall into the moderately positive range. These nations include Egypt, Brazil, Japan, South Korea, Mexico, Thailand, and the United States. The only countries in the negative range were India and the Dominican Republic. Moderate happiness also emerges in U.S. surveys of poor, physically disabled, unemployed, and elderly individuals. (For further information: Tracking global happiness. (1996) *Science News*. 149(24): 184.)

ANATOMY OF MELANCHOLY

In 1997, researchers found that one portion of the brain is significantly smaller and less active in people suffering from hereditary depression. In particular, researchers reported that they had zeroed in on a tiny, thimble-sized nodule of the brain located about 2.5 inches behind the bridge of the nose. Other scientists had already shown that this section of the brain, called the subgenual prefrontal cortex, plays an important role in the control of emotions. It can also be the trigger point for both bouts of paralyzing sadness and the wildly euphoric highs of bipolar disorder. It also exhibits signs of sluggish activity during periods of major depression and evidence of heightened activity during periods of manic euphoria. Depressed patients show almost 50 percent less brain tissue in the affected region. (For further information: Bower, B. (1997) Node emerges on brain's emotional network. *Science News*. 151: 254. Also see Gorman, C. (1997) Anatomy of melancholy. *Time*. May: 78.)

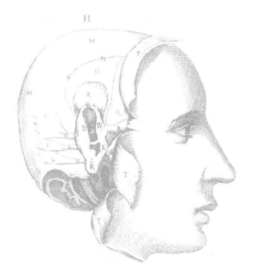

NOTES

PREFACE

1. The word *cryonics* was coined in 1965 and refers to freezing people for future revival. Today the Alcor Foundation, in Scottsdale, Arizona (info@alcor.org), uses cryonic technology for freezing a person after a terminal illness or a fatal accident, in the hope that medical science can revive that person in the future.

   Evidence for the possibility of biological resurrection continues to grow. For example, living animals can be "stopped" by placing them in suspended animation by freezing or dehydration. Later, the animals can be revived at the scientist's leisure. For example, *tardigrades* (small animals with four pairs of short, stubby legs armed with claws) can be dried out and then rehydrated like a backpacker's dinner. The creatures return from life after apparent death. In addition, thousands of healthy human babies have been born from frozen embryos.

   In the 1950s and 1960s, golden hamsters were cooled until 60 percent of their brain water was converted to ice and then thawed, demonstrating complete recovery with no behavioral abnormalities. (Smith, A.U. (1957) Studies on golden hamsters during cooling to and rewarming from body temperatures below 0 degrees centigrade. *Proc. Royal Society Biology London Series B*. 147: 517.) In the 1960s, isolated cat brains were treated with 15 percent glycerol and cooled to −20 degrees centigrade for five days. When warmed they returned to normal brain function as determined by EEG measurements. (Suda, I., and Kito, A.C. (1966) Histological cryoprotection of rat and rabbit brains. *Cryoletters* 5: 33.)

For more information on cryonics, contact: Alcor Foundation, 7895 E. Acoma Drive, #110, Scottsdale, Arizona 85260-6916. Readers interested in a newsletter published by a nonprofit organization promoting immortality, happiness, and cryonic preservation may contact *Venturist Monthly News*, 10444 N. Cave Creek Rd., Phoenix, Arizona 85020 (e-mail: mike@alcor.org). Their newsletter cites various cryonics organizations offering suspension services to the public, including: Alcor Life Extension Foundation (Scottsdale, Arizona), American Cryonics Society (Cupertino, California), CryoCare Foundation (Culver City, California), Cryonics Institute (Oak Park, Michigan), International Cryonics Foundation (Stockton, California), and TransTime Incorporated (Oakland, California).

2. The search for scientific ways to preserve the dead is not limited to recent cryonics efforts. In 1891, Dr. Varlot, a surgeon at a major hospital in Paris, developed a method for covering a person with a layer of metal in order to permanently preserve the person. Figure 1 illustrates how this is done with a child's cadaver. First the body is made electrically conductive by atomizing nitrate of silver on it. To free the silver in this solution, the corpse is placed under a glass dome from which the air is evacuated. Next the body is exposed to vapors of white phosphorus dissolved in carbon disulfide. Having been made conductive, the body is immersed in a galvanic bath of sulphate of copper, thus creating a millimeter thick layer of metallic copper deposited on the skin. The resulting body has a very strong, long-lasting, bright copper finish.

INTRODUCTION

1. Newton was not expected to survive his first day of life, much less 84 years. Deprived of a father before birth, he soon lost his mother as well.

Within two years of being widowed she remarried and left young Isaac with his grandmother to move to a neighboring village where she raised a son and two daughters. For nine years, until the death of his stepfather, Isaac was effectively separated from his mother. Some of his pronounced psychotic tendencies have been ascribed to this traumatic event. That he hated his stepfather we may be sure. When he examined the state of his soul in 1662 and compiled a catalog of sins, he remembered "threatening my father and mother Smith to burne them and the house over them." The acute sense of insecurity that rendered him obsessively anxious when his work was published, and irrationally violent when he defended it, accompanied Newton throughout his life.

CHAPTER 1

1. O'Neill, J. (1944) *Prodigal Genius*. David McKay Co.: New York, p. 64.
2. Tesla's biographers do not seem to mention whether these flashes of light have any connection with ocular migraines.
3. Cheney, M. (1981) *Tesla: Man Out of Time*. Dell: New York, p. 43.
4. Freudians would probably call some of Tesla's behavior anal retentive. In Freudian psychoanalytic theory, the anal stage is the period in a child's psychosexual development during which the child's main concerns are with the processes of defecation. Freudians believe this stage is significant for the child's later development because his responses to parental demands for bowel control may have far-reaching effects on his personality. Should a person become fixated, or locked, in the anal stage, he may develop what has been termed an anal character: frugal, orderly, and obstinate.

CHAPTER 2

1. Heaviside dealt with theoretical aspects of problems in telegraphy and electrical transmission, making use of an unusual calculatory method called operational calculus, now better known as the method of Laplace transforms, to study transient currents in networks. The use of operators in mathematics has a long history. Heaviside's great contribution was the application of operators to communication problems, but he did not teach mathematicians anything new. Rather he used operators

to reduce differential equations of a physical system to equivalent algebraic equations. This is what the Laplace transform does for the modern engineer. Heaviside was also the first to formulate the "telegrapher's equation" for voltage V as a function of distance x, time t, and resistance R in terms of capacitance and inductance, terms that he coined. The equation, whose coefficient could be optimized to produce a distortionless mode of propagation, proved to have wide application in general dynamics.

2. Freudians would probably call some of this behavior anal retentive— see note 4 for Chapter 1.

CHAPTER 7

1. Galton writes, "I had to hold a little court of justice on most days usually followed by corporal punishment, deftly administered. At a signal from me the culprit's legs were seized from behind, and he was thrown face forwards on the ground and held, while Hans applied the awarded number of whip strokes. This rough-and-ready justice became popular. Women, as usual, were the most common causes of quarrel."

CHAPTER 8

1. Attacks beginning with hallucinations most frequently originate in the lateral portion of the lobe; those beginning with automatic behavior, in the medial temporal structures.
2. This refers to Sir Frank Whittle (1907–1996), English inventor and aviator who produced the first British jet propulsion unit.

CHAPTER 9

1. As discussed later, child psychologist Ralph Meister gave six-year-old Kaczynski an IQ test that showed Kaczynski to be in the range of 160 to 170.
2. From a search of his Montana cabin, it appears Kaczynski was taking an antidepressant drug known as trazodone.
3. See, for example, the *New York Times* article cited in the reference section.

4. The *New York Times* reports his age as nine months. *Time* magazine reports his age as six months.

5. Gibbs et al. gives the address as "Khadar Khef." A quick check on the Internet shows one that Al-Khadar led a party in Egypt. Also, Khadar seems to be a common first name in the Middle East. It also appears in an Outer Earth playing card called the Moons of Khadar, which came out in summer 1996.

6. Mathematician Stanislaw Ulam estimated that mathematicians publish 200,000 theorems every year, and I have increased this figure to reflect the growing number of mathematical publications. The majority of these theorems are probably ignored, and only a minuscule fraction come to be understood and believed by any sizable group of mathematicians.

7. I do not know if his arrhythmia was merely an excuse for not dealing with family members, or if it resulted from hypochondria, or if it was a side effect from the antidepressant medication he may have been taking.

8. The May 26, 1996, issue of the *New York Times* reported that when his father was in the process of dying of lung cancer, Ted's brother marked the envelope properly with the red line, and Ted wrote back, "You have done well, this was something worth communicating." In either account, he had a quite distant relationship with his parents.

9. *Ice Brothers* was published by Arbor House, whose symbol is a tree leaf. We do not know if there is a connection between the tree leaf symbol and the victim's last name, "Wood." The book describes a Greenland patrol searching for German U-boats during World War II. The patrol used a boat named "The Valkyrie," which means "chooser of the fallen (or slain)." The book contained the following dedication: "Little remembered and little honored then, they'll never have to doubt that they were men." The author, Sloan Wilson, also wrote *The Man in the Gray Flannel Suit*, a "definitive epithet for the commuting-suburbanite, the status-hungry conformist from Madison Avenue."

10. This list of items and labels comes from a larger list compiled in the presence of Special Agent Tony Maxwell of the FBI and subsequently released to the public by order of Judge Charles Lovell in Helena.

11. The context for this statement is more clear from the complete paragraph from the Manifesto:

> "158. It presumably would be impractical for all people to have electrodes inserted in their heads so that they could be

controlled by the authorities. But the fact that human thoughts and feelings are so open to biological intervention shows that the problem of controlling human behavior is mainly a technical problem; a problem of neurons, hormones and complex molecules; the kind of problem that is accessible to scientific attack. Given the outstanding record of our society in solving technical problems, it is overwhelmingly probable that great advances will be made in the control of human behavior.

"159. Will public resistance prevent the introduction of technological control of human behavior? It certainly would if an attempt were made to introduce such control all at once. But since technological control will be introduced through a long sequence of small advances, there will be no rational and effective public resistance. . . .

"160. To those who think that all this sounds like science fiction, we point out that yesterday's science fiction is today's fact. The Industrial Revolution has radically altered man's environment and way of life, and it is only to be expected that as technology is increasingly applied to the human body and mind, man himself will be altered as radically as his environment and way of life have been."

CHAPTER 10

1. One patient-support organization to help people with obsessive–compulsive disorder is The O.C.D. Foundation, Inc, P.O. Box 9573, New Haven, Connecticut 06066. Another address is The O.C.D. Foundation, P.O. Box 70, Milford, Connecticut 06460-0070.

What triggers the brain to malfunction in people with obsessive–compulsive disorder? One recent article notes that people with the neurodegenerative disorder Huntington's disease can suffer classic signs of obsessive–compulsive disorder—a finding that interests researchers because the caudate nucleus in the basal ganglia is also damaged in this disease. In a few cases, a blow to the head has even been known to trigger obsessive–compulsive disorder. There is also increasing fascination in the idea that a mere bacterium can trigger this disorder. Infection with streptococcus normally causes nothing more than a sore throat. Some children, however, also develop a rare complication called Sydenham's chorea that changes the healthy child into one who suffers from compulsive behaviors. (For more information: Brown, P. (1997) Over and over. *New Scientist.* August 2, 2093: 27–31.)

2. About 60 scientific papers on trichotillomania have been published in the last five years, and there are also various electronic support groups on trichotillomania (TTM) that many have found useful. A good clearinghouse for information on TTM is: Trichotillomania Learning Center, 1215 Mission Street, Suite 2, Santa Cruz, California 95060. This group publishes a quarterly newsletter and has a board of directors consisting of researchers in the field. The Dean Foundation publishes a pamphlet on the disorder called: *Trichotillomania: A Guide.* For more information, write the Obsessive Compulsive Information Center, Dean Foundation for Health, Research, and Education, 8000 Excelsior Drive, Suite 302, Madison, Wisconsin 53717-1914. This pamphlet also lists some videos, nontechnical readings, and journal articles on the disorder.

CHAPTER 15

1. This quotation is from an anonymous eccentric in *Eccentrics, the Scientific Investigation* by D. Weeks and K. Ward (Stirling University Press).

CHAPTER 16

1. In the year 1534, Michelangelo's brother and father recently died, and Michelangelo was worried about his age and death. It was just at this time that the nearly 60-year-old artist wrote letters expressing strong feelings of attachment to young men. Some have interpreted this as evidence that Michelangelo was a homosexual, but such a reaction according to the artist's own statement would not be correct. No similar indications had emerged when the artist was younger. Many historians therefore suggest that these letters are consistent with the view that he was seeking a surrogate son.

2. Epilepsy is caused by abnormal electrical activity in the brain. Epilepsy includes generalized convulsions in which there is sudden unconsciousness with falling and shaking of limbs (grand mal seizure), momentary lapses of awareness (petit mal seizure), and local movements

and sensations in parts of the body (focal seizure), as well as other types of activity that may include strange automatic behavior, odd memories, hallucinatory experiences, and mood changes. Epilepsy is not a specific disease but rather a complex of symptoms that results from any of a number of conditions that excessively excite nerve cells of the brain. The terms cerebral seizure and convulsive disorder are often used interchangeably with epilepsy. Temporal lobe epilepsy is caused by abnormal electrical activity in the brain's temporal lobes located in the lower middle of the brain.

3. Neurologic abnormalities, such as epilepsy, can produce many symptoms resembling the psychoses. Psychoses are major mental illnesses that can cause major defects in judgment, delusions, hallucinations, defects in the thinking process, and the inability to objectively evaluate reality. It is difficult to differentiate psychoses from the class of less severe mental disorders known as psychoneuroses (commonly called neuroses) because a neurosis may be so severe that it actually constitutes a psychosis. Generally, patients suffering from the recognized psychotic illnesses exhibit an altered sense of reality and a disorganization of personality that sets them apart from neurotics. Such patients also often believe that nothing is wrong with them, despite the evidence to the contrary as revealed by their confused behavior. Psychoses may be divided into two categories: organic and functional. Organic psychoses are those caused by a known physical abnormality (e.g., organic brain disease). All other psychoses are called functional ones. Schizophrenia is usually categorized as a functional psychosis (even though one might easily debate that an organic cause for schizophrenia will become more apparent with continuing research.) Dementia is the principal syndrome in the most common and widespread organic psychosis, Alzheimer's disease.

4. Whitley Strieber remarks in *Communion*: "The corridor into our world could in a very sense be through our own minds. Maybe really skilled observation and genuine insight will cause the visitors to come bursting to the surface shaking like coelacanths in a net. Something is here, be it a message from the stars or from the booming labyrinth of the mind . . . or from both."

5. Schizophrenia affects about three million Americans and is as prevalent as the various forms of epilepsy. Schizophrenia involves delusions, hallucinations, and incoherent behavior. Some of the schizophrenic's behavior may also have epileptic roots. Like most psychiatric disorders (which were once thought to be caused by problems in parental up-

bringing), schizophrenia is now believed to have a viral, biological, and/or genetic cause. The brains of many schizophrenics show abnormality in the temporal lobes or temporolimbic region. Many schizophrenics have mothers who had the flu during the fifth month of pregnancy.

6. I view these legends with caution because I am not certain they could know about excess fluid at that time in history.

7. Why do the towering figures of genius seem to be more common in the past (Shakespeare, Newton, Michelangelo . . .) than today? Isn't the raw material for genius, for example talent and cultural opportunity, still alive today? The problem is one of complexity. Tesla did not need to master any special mathematics to excel. Pyke did not need to understand the molecular basis of materials. At least with physics, the answer is clear. In order to be great today, a modern physicist must master an amazing amount of mathematics and an expanding amount of previous knowledge. In order to make advances, the modern physicist must manipulate more theory and needs more mental power than required by scientists in past centuries. As James Gleick notes, consider for example the modern particle physicist who must master group theory, non-Abelian gauge theories, and the spin statistics and Yang–Mills.

APPENDIX A

1. Erdös pursued an incredibly wide range of mathematics and had outstanding results in a dozen or so different fields, some of which he created. He started as a number theorist. Some proofs from his early youth are still amazing to this day for their mathematical beauty. Later he grew interested in combinatorics, in particular to graph theory, and he became a leading scholar in probability theory. He also became one of the top authorities in classical set theory and founded new branches in partition theory. He also helped to found combinatorial geometry and transfinite combinatorics. He has literally thousands of theorems that are important in their respective disciplines. (For more information on Erdös, see Kolata G., Paul Erdös, a math wayfarer at Field's Pinnacle, dies at 83. *New York Times* (24 Sept, 1996). Also see MacKenzie, D., Homage to an itinerant master. (1997) *Science*. 275(5301): 759.)

2. Ludwig, A. (1995) *The Price of Greatness: Resolving the Creativity and Madness Controversy*. Guilford Press: New York, p. 90.

3. Darwin's angst. (1997) *Science*. 275(5299): 487.

4. Agoraphobia, the fear of being away from the safety of home, commonly occurs with panic disorder. Although the person avoids being alone, the only desired companions are those to whom no explanation of the agoraphobic behavior is required.

5. There is a close association between panic disorder and depression. A large percentage of people suffering from panic disorder go on to experience a major depression. Antidepressants such as imipramine and desipramine are the most effective treatment for panic disorder and may also provide effective relief of any associated depressive symptoms.

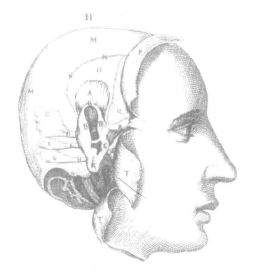

FURTHER READING

CHAPTER 1

Anderson, L. (1979) *Dr. Nikola Tesla Bibliography*. Ragusan Press: San Carlos, California.

Cheney, M. (1981) *Tesla: Man Out of Time*. Dell: New York.

Hall, S. (1986) Tesla: a scientific saint, wizard or carnival sideman? *Smithsonian*. June 71(11): 120.

Johnson, J. (1994) Extraordinary science and the strange legacy of Nikola Tesla. *Skeptical Inquirer*. Summer 18(4): 366–378.

O'Neill, J. (1944) *Prodigal Genius*. David McKay Co.: New York.

Popovich, V., Horvat, R. and Nikolich, N. (1956) *Nikola Tesla—Lectures, Patents, Articles*. Nikola Tesla Museum: Belgrade, Yugoslavia.

Seifer, M. (1996) *Wizard: The Life and Times of Nikola Tesla: Biography of a Genius*. Birch Lane Press: New York.

Tesla, N. and Childress, D. (1993) *The Fantastic Inventions of Nikola Tesla*. Adventures Unlimited Press: Kempton, Illinois.

Tesla, N. (1919) My inventions. *Electrical Experimenter.* May, June, July, October. Republished by Skolska Knjiga: Zagreb, Yugoslavia, 1977.

CHAPTER 2

Nahin, P. (1988) *Oliver Heaviside: Sage in Solitude.* IEEE Press: New York.
Nahin, P. (1990) Oliver Heaviside. *Scientific American.* June 262: 122–129.
Searle, G. (1987) *Oliver Heaviside, The Man.* Catt, I., ed. C.A.M. Publishing: St. Albans, England.

CHAPTER 4

Fallon, B.A., Liebowitz, M.R., Salman, E., Schneier, F.R., Jusino, C., Hollander, E. and Klein, D.F. (1993) Fluoxetine for hypochondriacal patients without major depression. *Journal of Clinical Psychopharmacology* 13(6): 438–441.
Somerville–Large, Peter (1990) *Irish Eccentrics.* Lilliput Press: Dublin.

CHAPTER 6

Berry, A. (1960) *Henry Cavendish: His Life and Scientific Work.* Hutchinson: London.
Wilson, G. (1851) *The Life of the Hon. Henry Cavendish ... and a Critical Inquiry into the Claims of All the Alleged Discoveries of the Composition of Water.* London.

CHAPTER 7

Galton, F. (1909) *Memories of My Life.* Methuen: London.
Gould, S. (1981) *The Mismeasure of Man.* Norton: New York.
Forrest, D. (1974) *Francis Galton: The Life and Work of a Victorian Genius.* Taplinger Publishing: New York. (The best biography on the subject.)

CHAPTER 8

Lampe, D. (1959) *Pyke, the Unknown Genius.* Evans Brothers Limited: London.

CHAPTER 9

Gibbs, N. (1996) Tracking down the unabomber. *Time*. April 147(16):38–46.
Gibbs, N., Lacayo, R., Morrow, L., Smolowe, J. and van Biema, D. (1996) *Mad Genius*. Warner Books: New York.
Lacayo, R. (1996) A tale of two brothers. *Time*. April 147(17):44–51.
McFadden, R. (1996) From a child of promise to the unabomb suspect. *New York Times*. Sunday, May 26, CXLV(50,439):1.

CHAPTER 10

The Editors of Time–Life Books (1992) *Odd and Eccentric People*. Time–Life: Alexandria, Virginia.
Michell, J. (1984) *Eccentric Lives and Peculiar Notions*. Citadel Press: Secaucus, New Jersey.
Neary, J. (1994) To the junk hound, life is one treasure hunt after another. *Smithsonian*. February 24(11): 79–84. (Discusses the question: Why does a sane man risk his wife's ire?)
Rapoport, J. (1989) *The Boy Who Couldn't Stop Washing*. Signet: New York.

CHAPTER 11

Holden, C. (1995) Brain trust. *Science*. March 267(5205):1764–1765.

CHAPTER 12

Adams, C. (1986) *The Straight Dope*. Ballantine: New York.
Diamond, M., Scheibel, A., Murphy, G. and Harvey, T. (1985) On the brain of a scientist: Albert Einstein. *Experimental Neurology*. 88: 198–204.
Kantha, S. (1992) Albert Einstein's dyslexia and the significance of Brodmann area 39 of his left cerebral cortex. *Medical Hypotheses*. 37: 119–122.
Roboz–Einstein, E. (1991) *Hans Albert Einstein: Reminiscences of His Life and Our Life Together*. Iowa Institute of Hydraulic Research: Iowa City.
Straus, E.G. (1982) Reminiscences. In *Albert Einstein: Historical and Cultural Perspectives*. Holton, G. and Elkana, Y., eds. Princeton University Press: Princeton, New Jersey; 417–423.

Wade, N. (1978) Brain that rocked physics rests in cider box. *Science*. 201: 696.

Wade, N. (1981) Brain of Einstein continues peregrinations. *Science*. 213: 521.

CHAPTER 13

Adams, C. (1986) *The Straight Dope*. Ballantine: New York.

Beyerstein, B. (1987–88) The brain and consciousness: Implications for psi phenomena. *Skeptical Inquirer*. Winter 12(2): 20–23.

Lorber, J. (1978) Is your brain really necessary? *Archives of Disease in Childhood*. 53(10): 834ff.

Lorber, J. (1981) Is your brain really necessary? *Nursing Mirror*. 152(18): 29–30.

vos Savant, M. (1992) *Ask Marilyn*. St. Martin's Press: New York.

CHAPTER 14

Bower, B. (1995) Criminal intellects: researchers look at why lawbreakers often brandish low IQs. *Science News*. April 147(15): 232–233.

Bower, B. (1995) IQ's evolutionary breakdown: intelligence may have more facets than testers realize. *Science News*. 147(14): 220–221.

Wu, C. (1995) Sometimes a bigger brain isn't better. *Science News*. August 148(8): 116.

CHAPTER 16

Bellodi, L., Scuito, G., Diaferia, G., Ronchi, P. and Smeraldi, E. (1992) A family history of obsessive–compulsive patients. *Clinical Neuropharmacology*. 15(1B): 308.

Blackmore, S. (1994) Alien abduction. *New Scientist*. Nov. 144(1952): 29–31.

Bower, B. (1995) Moods and the muse (a new study reappraises the link between creativity and mental illness). *Science News*. 147(24): 378–380.

Gutin, J. (1996) That fine madness. *Discover*. October 17(10): 75–82.

James, W. (1961) *The Varieties of Religious Experience*. Macmillan: New York.

Jamison, K. (1995) Manic–depressive illness and creativity. *Scientific American*. February 272(2): 62–67.

Jamison, K. (1993) *Touched with Fire: Manic–Depressive Illness and the Artistic Temperament*. The Free Press: New York.

LaPlante, E. (1993) *Seized*. HarperCollins: New York.

Ludwig, A. (1995) *The Price of Greatness: Resolving the Creativity and Madness Controversy*. Guilford Press: New York.

Mack, J. (1995) *Abduction* (revised edition). Ballantine: New York.

Robins, L.N. and Regier, D.A. (1991) *Psychiatric Disorders in America*. Free Press: New York.

Rocca, P., Maina, G., Ferrero, P., Bolgetto, F. and Raizza, L. (1992) Evidence of two subtypes of obsessive–compulsive disorder. *Clinical Neuropharmacology*. 15(1B): 315.

Strieber, W. (1987) *Communion*. Avon: New York.

Weeks, D. and Ward, K. (1988) *Eccentrics, the Scientific Investigation*. Stirling University Press: East Kilbridge, Scotland.

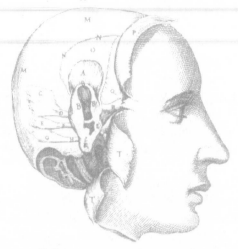

ABOUT THE AUTHOR

C lifford A. Pickover received his Ph.D. from Yale University's Department of Molecular Biophysics and Biochemistry. He graduated first in his class from Franklin and Marshall College, after completing the four-year undergraduate program in three years. He is the author of the popular books *Time—A Traveler's Guide* (Oxford University Press, 1998), *The Loom of God* (Plenum, 1997), *The Alien IQ Test* (Basic Books, 1997), *Black Holes—A Traveler's Guide* (Wiley, 1996), and *Keys to Infinity* (Wiley, 1995). He is also the author of *Chaos in Wonderland: Visual Adventures in a Fractal World* (1994), *Mazes for the Mind: Computers and the Unexpected* (1992), *Computers and the Imagination* (1991), and *Computers, Pattern, Chaos, and Beauty* (1990), all published by St. Martin's Press—as well as the author of over 200 articles concerning topics in science, art, and mathematics. He is also a coauthor, with Piers Anthony, of the science fiction novel *Spider Legs*.

Pickover is currently an associate editor for the scientific journals *Computers and Graphics*, *Computers in Physics*, and *Theta Mathematics Journal* and is an editorial board member for *Speculations in Science and Technology*, *Idealistic Studies*, *Leonardo*, and *YLEM*. He has been a guest editor for several scientific journals. Editor of *The Pattern Book: Fractals, Art, and Nature* (World Scientific, 1995), *Visions of the Future: Art, Technology, and Computing in the Next Century* (St. Martin's Press, 1993), *Future Health* (St. Martin's Press, 1995), *Fractal Horizons* (St. Martin's Press, 1996), and *Visualizing Biological Information* (World Scientific, 1995), and coeditor of the books *Spiral Symmetry* (World Scientific, 1992) and *Frontiers in Scientific Visualization* (Wiley, 1994), Pickover's primary interest is finding new ways to continually expand creativity by melding art, science, mathematics, and other seemingly disparate areas of human endeavor.

The *Los Angeles Times* recently proclaimed, "Pickover has published nearly a book a year in which he stretches the limits of computers, art and thought." Pickover received first prize in the Institute of Physics' "Beauty of Physics Photographic Competition." His computer graphics have been featured on the cover of many popular magazines, and his research has recently received considerable attention by the press—including CNN's "Science and Technology Week," The Discovery Channel, *Science News*, *The Washington Post*, *Wired*, and *The Christian Science Monitor*—and also in international exhibitions and museums. *Omni* magazine recently described him as "Van Leeuwenhoek's twentieth century equivalent." *Scientific American* several times featured his graphic work, calling it "strange and beautiful, stunningly realistic." Pickover has received U.S. Patent 5,095,302 for a 3-D computer mouse and 5,564,004 for strange computer icons.

Pickover is currently a Research Staff Member at the IBM T. J. Watson Research Center, where he has received 14 invention achievement awards, three research division awards, and four external honor awards. In addition to his other publications, Pickover is a lead columnist for the brain-boggler column in *Discover* magazine.

Dr. Pickover's hobbies include the practice of Ch'ang-Shih Tai-Chi Ch'uan (a form of martial arts) and Shaolin Kung Fu, raising golden and green severums (large tropical fish found in the central Amazon basin), and piano playing (mostly jazz). He is also a member of The SETI League, a worldwide group of radioastronomers and signal processing enthusiasts who systematically and scientifically search the heavens to detect evidence of intelligent extraterrestrial life. He can be reached at P.O. Box 549, Millwood, New York 10546-0549, USA. (Web site: http://sprott.physics.wisc.edu/pickover/home.htm)

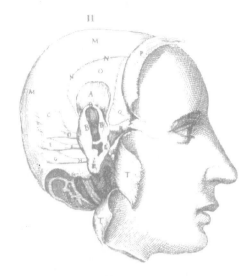

INDEX

Index